Primary SPACE Project Research Team

Research Coordinating Group

Professor Paul Black (Co-Director)
Dr. Jonathan Osborne

Professor Wynne Harlen (Co-Director)
Director, Scottish Council for Research in
Education
Professor Terry Russell (Deputy Director)

Centre for Educational Studies
King's College London
University of London
Cornwall House Annexe
Waterloo Road
London SE1 8TZ

Centre for Research in Primary Science
and Technology
Department of Education
University of Liverpool
126 Mount Pleasant
Liverpool L3 5SR

Tel: 0171 872 3094

Tel: 0151 794 3270

Project Researchers

Pamela Wadsworth (from 1989)

Derek Bell (from 1989)
Ken Longden (from 1989)
Adrian Hughes (1989)
Linda McGuigan (from 1989)
Dorothy Watt (1986-89)

Associated Researchers

John Meadows
(South Bank Polytechnic)

Bert Sorsby
John Entwistle
(Edge Hill College)

LEA Advisory Teachers

Maureen Smith (1986-89)
(ILEA)

Joan Boden
Karen Hartley
Kevin Cooney (1986-88)
(Knowsley)

Joyce Knaggs (1986-88)
Heather Scott (from 1989)
Ruth Morton (from 1989)
(Lancashire)

PRIMARY SPACE PROJECT
RESEARCH REPORT
January, 1991

Reprinted with corrections, 1998

Materials

by
TERRY RUSSELL, KEN LONGDEN
and LINDA McGUIGAN

LIVERPOOL UNIVERSITY PRESS

First published 1991 by
LIVERPOOL UNIVERSITY PRESS
Senate House, Abercromby Square
Liverpool, L69 3BX

Reprinted, with corrections, 1998

British Library Cataloguing-in-Publication Data
A British Library CIP record is available
ISBN 0-85323-147-8

Printed and bound in the European Union by
Redwood Books, Trowbridge, Wiltshire

CONTENTS

Page

INTRODUCTION ....

1.   **METHODOLOGY**   1

2.   **EXPLORATION**   5

3.   **CHILDREN'S IDEAS**   17

4.   **INTERVENTION**   59

5.   **EFFECTS OF INTERVENTION:**
        CHANGES IN CHILDREN'S IDEAS   95

6.   **SUMMARY**   139

        APPENDICES   145

ACKNOWLEDGEMENTS

The research reported here was made possible by the support of the Nuffield Foundation, the publishers Collins Educational and Lancashire Education Authority.

John Entwistle and Bert Sorsby are grateful for the support of Edge Hill College of Higher Education. The College provided a research grant and allowed them to take part in the early phase of the Project. Particular thanks are due to Bert Sorsby for the work on classification of materials.

For the word-processing, thanks are due to Jackie Bentley, Mary Cunningham, Dee Semple and Mary Stanton.

INTRODUCTION

This introduction provides an overview of the SPACE Project and its programme of research.

The Primary SPACE Project is a classroom-based research project which aims to establish:

* the ideas which primary school children have in particular science concept areas;

* the possibility of children modifying their ideas as the result of relevant experiences.

The research is funded by the Nuffield Foundation and the publishers, Collins Educational, and is being conducted at two centres, the Centre for Research in Primary Science and Technology, Department of Education, University of Liverpool and the Centre for Educational Studies, King's College, London. The joint directors are Professor Wynne Harlen and Professor Paul Black. The following local education authorities have been involved: Inner London Education Authority, Knowsley and Lancashire.

The Project is based on the view that children develop their ideas through the experiences they have. With this in mind, the Project has two main aims: firstly, to establish (through an elicitation phase) what specific ideas children have developed and what experiences might have led children to hold these views; and secondly, to see whether, within a normal class-room environment, it is possible to encourage a change in the ideas in a direction which will help children develop a more 'scientific' understanding of the topic (the intervention phase).

In the first phase of the Project from 1987 to 1989, eight concept areas were studied:

In the second phase of the Project during 1989 and 1990, a further ten concept areas were studied:

Electricity	Earth
Evaporation and condensation	Earth in space
Everyday changes in non-living materials	Energy
	Genetics and evolution
Forces and their effect on movement	Human influences on the Earth
	Processes of life
Growth	Seasonal changes
Light	Types and uses of materials
Living things' sensitivity to their environment	Variety of life
	Weather
Sound	

Research reports are usually based on each of these concept areas; occasionally, where the areas are closely linked, they have been combined in a single report.

The Project has been run collaboratively between the university research teams, local education authorities and schools, with the participating teachers playing an active role in the development of the Project work.

Over the life-span of the Project a close relationship has been established between the university researchers and teachers, resulting in the development of techniques which advance both classroom practice and research. These methods provide opportunities, within the classroom, for children to express their ideas and to develop their thinking with the guidance of the teacher, and also help researchers towards a better understanding of children's thinking.

The Involvement of the Teachers

Schools and teachers were not selected for the Project on the basis of a particular background or expertise in primary science. In the majority of cases, two teachers per school were involved. This was advantageous in providing mutual support. Where possible, the authority provided supply cover for the teachers so that they could attend Project sessions for preparation, training and discussion during the school day. Sessions were also held in the teachers' own time, after school.

The Project team aimed to have as much contact as possible with the teachers throughout the work to facilitate the provision of both training and support. The diversity of experience and differences in teaching style which the teachers brought with them to the Project meant that achieving a uniform style of presentation in all classrooms would not have been possible, or even desirable. Teachers were encouraged to incorporate the Project work into their existing classroom organisation so that both they and the children were as much at ease with the work as with any other classroom experience.

The Involvement of Children

The Project involved a cross-section of classes of children throughout the primary age range. A large component of the Project work was classroom-based, and all of the children in the participating classes were involved as far as possible. Small groups of children and individuals were selected for additional activities or interviews to facilitate more detailed discussion of their thinking.

The Structure of the Project

In the first phase of the Project, for each of the topic areas studied, a list of concepts was compiled to be used by researchers as the basis for the development of work in that area. These lists were drawn up from the standpoint of accepted scientific understanding and contained concepts which were considered to be a necessary part of a scientific understanding of each topic. The lists were not necessarily considered to be statements of the understanding which would be desirable in a child at age eleven, at the end of the primary phase of schooling. The concept lists defined and outlined the area of interest for each of the studies; what ideas children were able to develop was a matter for empirical investigation.

In the second phase of the Project, the delineation of the topic area was informed by the National Curriculum for Science in England and Wales. The topic area was broken into a number of themes from which issues were selected for research. Themes sometimes contained a number of interlocking concepts; in other instances, they reflected only one under lying principle.

Most of the Project research work can be regarded as being organised into two major phases, each followed by the collection of structured data about children's ideas. These phases, called 'Exploration' and 'Intervention', are described in the following paragraphs and, together with the data collection, produce the following pattern for the research.

Phase 1a	Exploration
Phase 1b	Pre-Intervention Elicitation
Phase 2a	Intervention
Phase 2b	Post-Intervention Elicitation

The Phases of the Research

For the first eight concept areas, the above phases were preceded by an extensive pilot phase. Each phase, particularly the pilot work, was regarded as developmental; techniques and procedures were modified in the light of experience. The modifications involved a refinement of both the exposure materials and the techniques used to elicit ideas. This flexibility allowed the Project team to respond to unexpected situations and to incorporate useful developments into the programme.

Pilot Phase

There were three main aims of the pilot phase. They were, firstly, to trial the techniques used to establish children's ideas; secondly, to establish the range of ideas held by primary school children; and thirdly, to familiarise the teachers with the classroom techniques being employed by the Project. This third aim was very important since teachers were being asked to operate in a manner which, to many of them, was very different from their usual style. By allowing teachers a 'practice run', their initial apprehensions were reduced, and the Project rationale became more familiar. In other words, teachers were being given the opportunity to incorporate Project techniques into their teaching, rather than having them imposed upon them.

Once teachers had become used to the SPACE way of working, a pilot phase was no longer essential and it was not always used when tackling the second group of concept areas. Moreover, teachers had become familiar with both research methodology and classroom techniques, having been involved in both of them. The pace of research could thus be quickened. Whereas pilot, exploration and intervention had extended over two or three terms, research in each concept area was now reduced to a single term.

In the exploration phase children engaged with activities set up in the classroom for them to use, without any direct teaching. The activities were designed to ensure that a range of fairly common experiences (with which children might well be familiar from their everyday lives) was uniformly accessible to all children to provide a focus for their thoughts. In this way, the classroom activities were to help children articulate existing ideas rather than to provide them with novel experiences which would need to be interpreted.

Each of the topics studied raised some unique issues of technique and these distinctions led to the exploration phase receiving differential emphasis. Topics in which the central concepts involved long-term, gradual changes, such as 'Growth', necessitated the incorporation of a lengthy exposure period in the study. A much shorter period of exposure, directly prior to elicitation, was used with topics such as 'Light' and 'Electricity' which involve 'instant' changes.

During the exploration, teachers were encouraged to collect their children's ideas using informal classroom techniques. These techniques were:

(a)   **Using Log-Books (free writing/drawing)**
        Where the concept area involved long-term changes, it was suggested that children should make regular observations of the materials, with the frequency of these depending on the rate of change. The log-books could be pictorial or written, depending on the age of the children involved, and any entries could be supplemented by teacher comment if the children's thoughts needed explaining more fully. The main purposes of these log-books were to focus attention on the activities and to provide an informal record of the children's observations and ideas.

(b)   **Structured Writing/Annotated Drawing**
        Writing or drawings produced in response to a particular question were extremely informative. Drawings and diagrams were particularly revealing when children added their own words to them. The annotation helped to clarify the ideas that a drawing represented.

        Teachers also asked children to clarify their diagrams and themselves added explanatory notes and comments where necessary, after seeking clarification from children.

        Teachers were encouraged to note down any comments which emerged during dialogue, rather than ask children to write them down themselves. It was felt that this technique would remove a pressure from children which might otherwise have inhibited the expression of their thoughts.

(c) **Completing a Picture**

Children were asked to add the relevant points to a picture. This technique ensured that children answered the questions posed by the Project team and reduced the possible effects of competence in drawing skills on ease of expression of ideas. The structured drawings provided valuable opportunities for teachers to talk to individual children and to build up a picture of each child's understanding.

(d) **Individual Discussion**

It was suggested that teachers use an open-ended questioning style with their children. The value of listening to what children said, and of respecting their responses, was emphasised as was the importance of clarifying the meaning of words children used. This style of questioning caused some teachers to be concerned that, by accepting any response whether right or wrong, they might implicitly be reinforcing incorrect ideas. The notion of ideas being acceptable and yet provisional until tested was at the heart of the Project. Where this philosophy was a novelty, some conflict was understandable.

In the elicitation which followed the exploration phase, the Project team collected structured data through individual interviews and work with small groups. The individual interviews were held with a random, stratified sample of children to establish the frequencies of ideas held. The same sample of children was interviewed pre- and post-intervention so that any shifts in ideas could be identified.

Intervention Phase

The elicitation phase produced a wealth of different ideas from children, and produced some tentative insights into experiences which could have led to the genesis of some of these ideas. During the intervention, teachers used this information as a starting point for classroom activities, or interventions, which were intended to lead to children extending their ideas. In schools where a significant level of teacher involvement was possible, teachers were provided with a general framework to guide their structuring of classroom activities appropriate to their class. Where opportunities for exposing teachers to Project techniques had been more limited, teachers were given a package of activities which had been developed by the Project team.

Both the framework and the intervention activities were developed as a result of preliminary analysis of the pre-intervention elicitation data. The intervention strategies were:

(a) **Encouraging children to test their ideas**

It was felt that, if pupils were provided with the opportunity to test their ideas in a scientific way, they might find some of their ideas to be unsatisfying. This might encourage the children to develop their thinking in a way compatible with greater scientific competence.

(b) **Encouraging children to develop more specific definitions for particular key words**
Teachers asked children to make collections of objects which exemplified particular words, thus enabling children to define words in a relevant context, through using them.

(c) **Encouraging children to generalise from one specific context to others through discussion**
Many ideas which children held appeared to be context-specific. Teachers provided children with opportunities to share ideas and experiences so that they might be enabled to broaden the range of contexts in which their ideas applied.

(d) **Finding ways to make imperceptible changes perceptible**
Long-term, gradual changes in objects which could not readily be perceived were problematic for many children. Teachers endeavoured to find appropriate ways of making these changes perceptible. For example, the fact that a liquid could 'disappear' visually yet still be sensed by the sense of smell - as in the case of perfume - might make the concept of evaporation more accessible to children.

(e) **Testing the 'right' idea alongside the children's own ideas**
Children were given activities which involved solving a problem. To complete the activity, a scientific idea had to be applied correctly, thus challenging the child's notion. This confrontation might help children to develop a more scientific idea.

(f) **Using secondary sources**
In many cases, ideas were not testable by direct practical investigation. It was, however, possible for children's ideas to be turned into enquiries which could be directed at books or other secondary sources of information.

(g) **Discussion with others**
The exchange of ideas with others could encourage individuals to reconsider their own ideas. Teachers were encouraged to provide contexts in which children could share and compare their ideas.

In the post-intervention elicitation phase the Project team collected a complementary set of data to that from the pre-intervention elicitation by re-interviewing the same sample of children. The data were analysed to identify changes in ideas across the sample as a whole and also in individual children.

These phases of Project work form a coherent package which provides opportunities for children to explore and develop their scientific understanding as a part of classroom activity, and enable researchers to come nearer to establishing what conceptual development it is possible to encourage within the classroom and the most effective strategies for its encouragement.

The Implications of the Research

The SPACE Project has developed a programme which has raised many issues in addition to those of identifying and changing children's ideas in a classroom context. The question of teacher and pupil involvement in such work has become an important part of the Project, and the acknowledgement of the complex interactions inherent in the classroom has led to findings which report changes in teacher and pupil attitudes as well as in ideas. Consequently, the central core of activity, with its data collection to establish changes in ideas, should be viewed as just one of the several kinds of change upon which the efficacy of the Project must be judged.

The following pages provide a detailed account of the development of the 'Materials' topic, the Project findings and the implications which they raise for science education.

The research reported in this and the companion research reports, as well as being of intrinsic interest, informed the writing and development with teachers of the Primary SPACE Project curriculum materials, published by Collins Educational.

1. METHODOLOGY

Most of the research reported in this volume was carried out in six Lancashire schools between January and May 1990. The sample, research programme and defining of the topic are described in sections 1.1 to 1.3. A study of children's classification of materials is also reported. This research was carried out in six different schools, also in Lancashire, during 1987. It is described in section 1.4.

1.1   Sample

a)   Schools

The six participating schools belong to Lancashire Local Education Authority and are situated in the region of Preston and Ormskirk. They include rural and urban schools. Five of the schools contain the whole primary range while the sixth is a junior school. The classes included in the sample range from Reception (usual age five) to Year Six (usual age eleven) with some mixed-age classes.

Names of schools, headteachers and teachers are given in Appendix 1.

b)   Teachers

Twelve teachers were involved in the work of the Project. Nine of them had previous experience of SPACE; five had participated from the beginning of the Project.

A general principle was that, where possible, each school should have two participating teachers for the purpose of mutual support; in one school, however, three teachers were involved and in another, one teacher worked alone.

Teachers received further support through two whole-group meetings, each scheduled for two hours in an afternoon. At these meetings, strategies for exploring and developing children's ideas were discussed and planned. Visits to the classes were also made by the Project team including an advisory teacher.

Most of the teachers were aware of SPACE philosophy and techniques from their work on other topics within the Project. The three teachers new to the Project were informed about the SPACE approach; they received extra support from the university researchers when this was requested.

c)   Children

All children in the twelve classes were involved in the Project work to some extent. A stratified random sample was selected to be interviewed by members of the Project team at two stages during the work. Teachers were asked to assign each child in their class to an achievement band (high, medium, low) related to their overall school performance. Children were then randomly selected from the class lists so that numbers were balanced by achieve-

2

ment band and gender. This same sample was used by some teachers in cases where data collection necessitated one-to-one work between child and teacher. In most cases, however, data were collected from all the children.

1.2   The Research Programme

Classroom work took place in two major phases. The first of these was called 'exploration'. It had two interconnected aspects. One was to expose children to the topic (of materials); the other was to elicit their ideas in relation to that topic.

The first aspect, exposure, stems from the intention that the second, elicitation, should occur after children had had some opportunity to explore and handle materials. By setting the context using real materials, it was hoped to collect a more considered view rather than catching children by surprise.

The exploration of children's ideas by the class teacher was followed by interviews with a random sample of the class. These interviews were conducted by researchers and the advisory teacher.

The second major phase was termed 'intervention'. During the previous phase, teachers had been 'holding back' to allow children to express their own ideas. In the intervention, teachers offered children experiences which gave them an opportunity to reflect on their ideas, test them out, discuss them and amend, reject or retain them.

This was followed by a limited post-intervention elicitation for the whole class coupled with interviews of the same sub-sample as before. The research was carried out between January and May 1990. The sequence of events is given in figure 1.1:

**Figure 1.1:
Project sequence
of events**

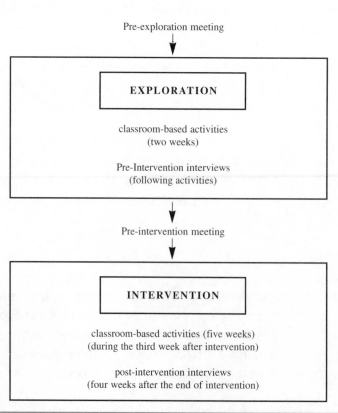

Pre-exploration meeting

EXPLORATION

classroom-based activities
(two weeks)

Pre-Intervention interviews
(following activities)

Pre-intervention meeting

INTERVENTION

classroom-based activities (five weeks)
(during the third week after intervention)

post-intervention interviews
(four weeks after the end of intervention)

1.3   Defining the Topic

1.3.1   Division into themes

In previous SPACE work, a list of concepts was drawn up for each topic. This list was intended to delineate the boundaries within which the research would be carried out. With the advent of a National Curriculum for England and Wales, aspects of the document for Science provided a framework in which concepts for investigation could be discerned (reference: Department of Education Science and the Welsh Office (1989), *Science in the National Curriculum*, H.M.S.O., London).

The National Curriculum defines programmes of study (PoS) for different age groups of children. The programmes of study indicate the kinds of experiences to which all children of that age group are expected to be exposed. For the primary age range of children, two programmes of study are relevant, that for Key Stage One (KS1) being applicable for children up to the school year in which they have their seventh birthday (infants) and that for Key Stage Two (KS2) being applicable for children up to the school year in which they have their eleventh birthday (Juniors).

At the time, the National Curriculum also defined statements of attainment (SoA) applicable throughout the years of compulsory education. Statements of attainment were written for each of a series of ten levels. They indicated what a child must 'know', 'understand', or 'be able to do', in order for some attainment to be assigned to that level. The levels most likely to be achieved by primary aged children are level one to level five.

Each subject of the National Curriculum is divided into a number of Attainment Targets (AT). Each target reflects a particular aspect of the subject which is deemed worthy of separate assessment. There are seventeen such targets for Science and AT6, 'Types and Uses of Materials', pertains to the topic of this report, 'Materials'.

This Attainment Target was taken and itself divided into themes, each of which seemed to reflect a relatively discrete concept area within the topic of materials. Both programmes of study and statements of attainment for levels one to five were taken into account when making this sub-division. The outcome is shown as Figure 1.2 Programmes of study are preceded by 'KS1' or 'KS2' while statements of attainment are prefixed by a number denoting the level and a letter indicating its position within the level. Thus, statement 4c is the third statement at level four.

Five themes were defined by SPACE researchers, although these do not carry equal weight. Theme D, here called 'Describing and Comparing Materials', is a strong central theme which pervades the whole topic of materials. Indeed, this makes it difficult to split up the programmes of study, and rather than take out short phrases which pertain to themes A, B and C, these have been retained within the major themes, D and E.

One school term was available for this research. It was thus not possible to tackle all five themes rigorously. Theme A relates to SPACE research already carried out in connection with AT13 'Energy' and to work previously published in the SPACE research report

'Evaporation and Condensation'. Theme C contains statements only at level five. It was therefore decided to focus on themes B, D and E.

The previous SPACE research into 'Change in materials' (which included classification of materials by children) further limited the extent of research required.

1.3.2 Issues within the themes

The following issues were chosen from within themes B, D and E to form the focus of the research:

1. *Descriptions and properties of materials*
   How do children describe materials? What properties do they make use of in their descriptions? What properties do they understand and/or make use of when interacting with materials?

2. *Solids, liquids and gases (states of matter)*
   What concepts of solid, liquid and gas do children hold?

3. *Uses of materials*
   What ideas do children have about why particular materials are chosen for particular purposes?

4. *Origins and manufacture of materials and change in materials*
   To what extent are children aware of where materials come from and the processes materials undergo during extraction and conversion to manufactured products? To what extent are children aware of the potential of various materials for change into other forms?

1.4 Research on Children's Classification of Materials

Some research into how children classified materials was carried out during 1987 as part of a study of children's ideas about change in materials. Six schools in the Lancashire Local Education Authority were involved. These schools, together with the names of the headteachers and participating teachers, are listed in Appendix II. They are situated in the region of Ormskirk and Skelmersdale, and include both rural and urban schools.

This research took place in the first phase of the SPACE Project. The eleven teachers were all volunteers and, in general, had no particular expertise in science.

In all but one of the schools, two teachers were working with the SPACE Project. This gave an opportunity for discussion about the work involved and for mutual support. Further support was provided by researchers.

The sample from which data were collected is described in section 3.1.1. of the report.

2. EXPLORATION

2.1   Design of Exploration Experiences and Interview Questions

This first stage of classroom-based research was exploratory in two senses. It was an exploration of children's thinking by the Project team. It was also an opportunity for children to start exploration of their own ideas. A sequence of activities was thus needed with the dual function of:

1.   *exposure* of children to the concept area of materials;

2.   *elicitation* of children's ideas.

This first aspect, exposure, reflects the Project's philosophy that elicitation should not be sprung on children. Through handling materials children would have an opportunity to make clear their own thinking to themselves. Any view elicited would thus be more likely to be a considered one rather than one made up on the spur of the moment. Although it is true that such 'exposure' could cause reconsideration and change in ideas, this does not interfere with the Project's aim of obtaining a 'snapshot' of the kind of views that children hold about materials.

Four focal issues were mentioned in Chapter One. These were, firstly, *descriptions and properties of materials;* secondly, *states of matter;* thirdly, *uses of materials;* and fourthly, *origins, manufacture and change in materials.* It was hoped to cover all four of these issues by combining information from classroom activities with that from structured interviews. The content of those activities was thus influenced by consideration of which aspects of the issues most suited classroom procedure and which appeared to be better approached through an interview question.

The exact design of the activities was greatly influenced by knowledge and expertise gleaned from previous SPACE research. Most of the teachers were familiar with the approach. Both they and research staff had become aware, from previous experience, of the kinds of activities that had proved successful in meeting the dual functions of 'exposure' and 'elicitation'. Such activities gave children a chance to think and also drew out their considered opinions. The successful activities had been based on certain techniques such as the use of open questions by the teacher in discussion with children. (Other techniques suitable for an exploration phase are described in the introduction to this report). More fundamentally, teachers were aware of the rationale for adopting those techniques. The techniques enabled them to establish children's ideas and it was those ideas which would serve as starting points in subsequent learning. This required teachers to adopt a role in which they deliberately held back from guiding children's thinking. They were to help children to clarify their ideas rather than seeking to make them justify and reconsider them.

6

Describing and comparing materials

KS1 Children should collect and find similarities and differences in a variety of everyday materials, natural and manufactured, including cooking ingredients, rocks, air, water and other liquids.
They should work with and change some of these materials by simple processes such as dissolving, squashing, pouring, bending, twisting and treating surfaces.

KS2 Children should work with a number of different everyday materials grouping them according to their characteristics, similarities and differences. Properties such as mass ('weight'), volume, strength, hardness, flexibility, compressibility and solubility should be investigated and measured.
Children should explore chemical change in a number of everyday materials, such as mixing Plaster of Paris, making concrete and firing clay.

1a be able to describe familiar and unfamiliar objects in terms of simple properties, for example, shape, colour, texture, and describe how they behave when they are, for example, squashed and stretched.

2a be able to recognise important similarities and differences including hardness, flexibility and transparency, in the characteristics of materials.

2b be able to group materials according to their characteristics.

3b be able to list the similarities and differences in a variety of everyday materials.

4a be able to make comparisons between materials on the basis of simple properties, strength, hardness, flexibility and solubility.

linked in that properties of materials determine their uses

(D)

(E)

Origins and uses of materials

KS2 Using secondary sources, they should explore their origins and how materials are used in construction. They should find out the common use of materials and relate the use of the properties which they have investigated, such as changes brought about by heating and cooling. They should learn about the dangers associated with the use of everyday materials, such as bleach and hot oil.

3a know that some materials occur naturally while many are made from raw materials.

4b be able to relate knowledge of these properties to the everyday use of materials.

Types and Uses of Materials

Heating and cooling materials

(A)

2c know that heating and cooling materials can cause them to melt or solidify or change permanently.

4d understand the sequence of changes of state that results from heating or cooling.

subthemes strongly linked

(B)

Solids, liquids, gases

4c know that solids and liquids have 'weight' which can be measured and, also, occupy a definite volume which can be measured.

4c be able to classify materials into solids, liquids and gases on the basis of their properties.

5a know that gases have 'weight'.

(C)

Simple 'chemistry'

5b be able to classify aqueous solutions as acidic, alkaline or neutral, by using indicators.

5c be able to give an account of the various techniques for separating and purifying mixtures.

Figure 1.2 Themes within Attainment Target 6

Previous research was another influence on the design of activities. A very brief outline of some of that research follows. The references are given in the Bibliography which appears as Appendix VIII.

Previous research into children's ideas about materials

It was during the second wave of Piaget's experimental research that he investigated children's ideas about conservation of matter. These studies were first reported in book form in French in 1941 and translated into English in 1974 (Piaget and Inhelder [1974]). These now-classical studies involved asking children whether they thought amounts, weight and volume had changed after transformation of a piece of material. Thus, for example, a ball of clay might be flattened or stretched into a sausage and a child asked if there was the same amount of clay after as there had been before. In the same book are also reported studies on conservation during dissolving and on the development of conceptions of density and compression/decompression. Piaget and Inhelder argued that it is through developing an underlying particle schema about the nature of matter that the child comes to believe in conservation during transformations of form (e.g. flattening, stretching, cutting) and state (e.g. dissolving).

This research spawned a great number of similar studies (see, for example, Modgil ([1974]) pp. 63-71). Those working within the constructivist philosophy of learning have also examined conservation both for physical changes such as changes of state and dissolving and for chemical change such as burning. In one of these studies, Andersson (1984) found that only about a third of a sample of fifteen-year-olds conserved mass when some sugar was dissolved. Driver (1985) has provided an overview of this whole area of study. This includes an analysis of the interaction between perceptually salient features of an event and interpretations in terms of the particulate nature of matter. She also pointed out that particle ideas do not guarantee conservation since individual particles may themselves be believed to change, to shrink under certain circumstances, for example. These constructivist studies also go beyond conservation, in that they look at the range of ideas that children might have about changes to materials.

Much of the work with children at a secondary level of education has focused on interpretation of events in particle terms (see, for example, Brook et al. [1984]). Some work with younger children has addressed the same issue (Novick and Nussbaum [1981]). It is still not clear, however, whether Piaget's claim of the spontaneous generation of atomistic ideas (or at least their formation prior to a belief in conservation) can be supported.

Other work with younger children has examined how those children actually interpret the changes that they observe. Osborne and Cosgrove (1983) reported on the work of the Learning in Science Project of New Zealand in this respect. Children were shown instances of boiling, condensing, evaporating, dissolving and melting and were asked about what they thought was happening. Children's interpretations of a number of instances of one of these changes, evaporation, are discussed by Russell et al. (1989) (fully reported in Russell and Watt [1990]).

Changes such as evaporation transform the state of a substance. Some research has been carried out on the meanings children attach to the names of those states. Stavy and Stachel (1985) introduced children to the terms 'solid' and 'liquid', asked them to classify a set of materials and then to predict which materials could be 'piled up'. They report the correct classification of liquids from an early age due to the idea that 'all liquids are made of water' whereas some kinds of solid were less well classified. Similarly Jones et al. (1989) asked children to categorise one set of substances as solid or not and another as liquid or not. They noted a considerable variation in children's responses according to the substances in question.

Notions of the gaseous state held by children of eleven to thirteen years have been investigated by Séré (1985). These children associated air with movement. This relates to the findings of Piaget (1930). He reported that young children appeared only to recognise air in movement, that is, as wind.

It can be seen from this account that conservation, change of state and nature of state have been extensively researched. Where any of these three areas of study was touched upon in the present research, care was taken to complement rather than duplicate. Only the third area, the nature of the states solid, liquid and gas, was examined extensively in this present study. This involved drawing out children's views in as open a way as possible in line with the general philosophy of the SPACE Project.

In contrast, other key areas in the field of materials - typical physical properties (like hardness, strength, transparency), the uses and origins of materials - do not appear to have been extensively investigated elsewhere.

2.2 Exploration Experiences - exposure and elicitation

At a pre-exploration meeting in January 1990 the team of teachers, advisory teacher and university personnel discussed ideas for exploration activities. The topic was outlined with the aid of Figure 1.2 (see Chapter One) which divided the Attainment Target 'Types and Uses of Materials' into themes. Some suggestions for exploration experiences were made.

These were based on the considerations outlined in the previous section. That is, both previous research and the techniques developed during SPACE research influenced the selection of possible activities. The suggestions were modified, clarified and added to as teachers discussed in groups whether and how the ideas might be tackled in their own classrooms. The outcome of the meeting was an agenda for a sequence of exploration experiences. This was written up after the meeting and sent to teachers. It appears in full form as Appendix III.

Participating children's ages ranged from five to eleven. The experiences were therefore phrased in as non-age-specific a way as possible. To some extent, they represent a compromise between the approach most suitable for a teacher of infants and that for a teacher of older children. Teachers were aware of this and were able to adapt to their own situations what was presented as a common framework. In particular, teachers of younger children with limited writing skills had to rely more heavily on drawings and oral comments. These techniques were, of course, also available to the other teachers, but generally they did not have to bear the same management demands of reporting data on behalf of the child as did the teachers of infants. It was thus necessary to provide a greater degree of classroom support to some teachers in order to free them for recording children's ideas.

It should also be clear from the time constraints of the Project that individual concepts within this particular topic could not be covered in depth. Some earlier SPACE research was able to focus on a small number of concepts. The two related concepts of evaporation and condensation, for instance, were the focus of one study and have already been published as a SPACE research report. Materials is a large concept field in which a number of concepts are embedded (including those of evaporation and condensation). Figure 2.1 gives an indication of some of the underlying ideas and their interrelation.

Since not every property of materials could be addressed, it was necessary to select while, at the same time, trying to ensure coverage of the main issues that had been identified.

The exploration phase experiences had the following titles:

Display

Comments on display materials. (Activity 1)

Finding out about materials. (Activity 2)

Where does it come from? (Activity 3)

Comparing materials. (Activity 4)

Ideas about solid, liquid, gas. (Activity 5)

Each of these items is discussed in turn.

The display and the first two activities which linked closely to it had a high degree of 'exposure' function. Teachers were thus advised to start with the display.

The display was a collection of everyday materials which quickly served to arouse interest and ensure engagement with the topic. These included foodstuffs and other items found in a kitchen, plastic and metal objects, cloth, building materials and other household items such as paint, oil and polish. Both natural and manufactured items were represented. Similarly there were examples of the three states of matter. The materials were not organised in any particular way. They were kept in their usual containers but other containers were available so that children could look more closely at those materials that were hidden by their packaging. Materials were labelled if it was not obvious what they were. Children were encouraged to look at the display. A number of questions placed in proximity to the collection prompted children to think about what they saw. Teachers were aware of the safety implica-

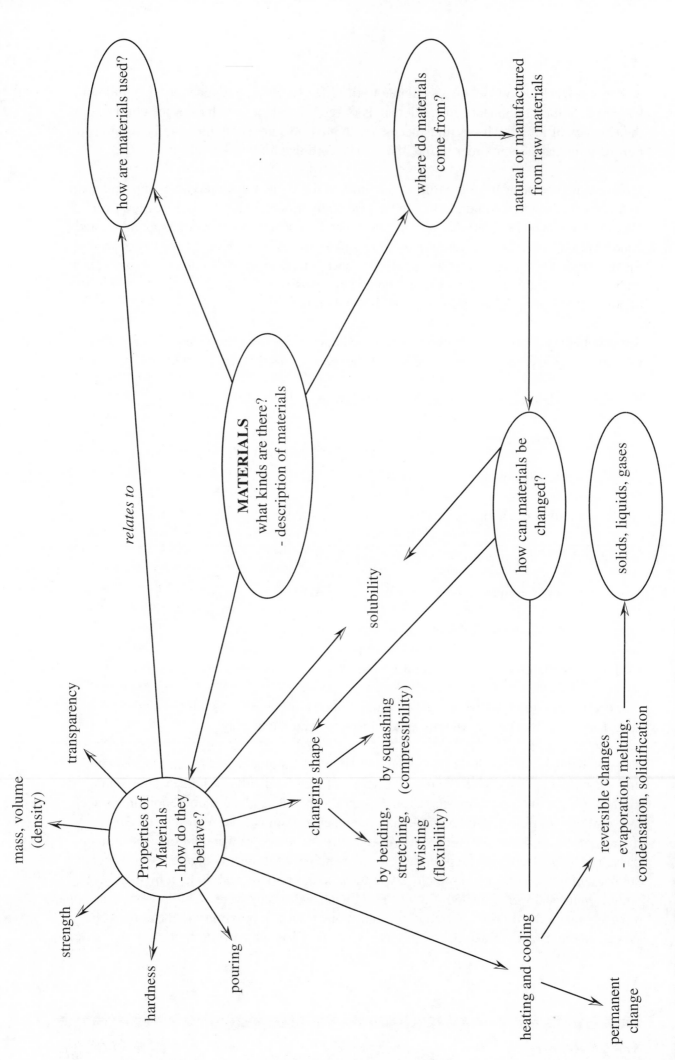

Figure 2.1 Some underlying ideas in 'types and uses of materials'

tions of a display of this kind and warned children of the dangers of smelling, tasting and touching unknown substances.

The first two activities developed further involvement with the display. The first of these, 'Comments on display materials', encouraged children to record their own observations or recollections about any of the items. They wrote down (or had written down for them) what they could say about items of their choice. They recorded the material, what they could say about it and the date on a sheet of paper. Children's individual record sheets could be added to as they wished.

The second activity, 'Finding out about materials', went beyond observation and description by allowing children to interact with the materials and test them in various ways. Teachers made available various 'implements', judging what was suitable for the age group concerned. The examples suggested were water and suitable containers, file, pliers, torch, magnet, sandpaper, sieve, pestle and mortar (or alternative), scissors, hammer and nail, magnifying glass, moulds. Children were asked to select one implement and a range of materials and to find out what they could about those materials. Teachers were advised to guide children towards the selection of appropriate materials, if necessary. Children (or the teacher) made a record of what they did and found out. These investigations were very much of a preliminary nature. The teacher did not intervene to encourage rigorous and fair testing. Rather the intention was to reveal what properties children thought salient and what they thought they were testing with the 'implements'.

These two activities both addressed the first of the four issues, that of description of materials and awareness of their properties. For the first activity, the interest was in the ways children talked or wrote about materials. They might describe what they observed or they might use recalled knowledge about the materials. In the second activity it was of interest to see what properties children revealed an awareness of or what property they might be testing when using a particular implement. In general, these two activities might indicate what properties of materials children focused on and showed an understanding of.

The third activity, 'Where does it come from?', required children to give their ideas about the origins of materials. Three different materials were chosen: metal in the form of a spoon, cotton in the form of some coloured fabric, and flour. Two of these, the cotton and the flour, are organic materials while the third, metal, has been manufactured from a raw material, its ore. All three materials have, however, been subject to a number of processes to produce the final form. For each object, children were asked to make annotated drawings showing what it was like, what it was like before that - and before that - and so on, as far back as they could go. To make this procedure clear, a familiar example was worked through with children. This activity addressed the fourth issue, that of the origins of materials and the extent that children are aware of the changes that take place during the manufacture of a particular item.

The fourth activity, 'Comparing materials', like the first two, addressed the issue of how children describe materials and what properties they make use of in their descriptions. Children were asked to describe materials, this time in the context of making a comparison. The activity thus focused on the properties that children treated as salient when pointing out similarities and differences. Schools were provided with five identical, transparent, sample containers, each one labelled and containing a different material. These were a piece of steel rod, a piece of cotton wool, some treacle, some talcum powder and some malt (coloured) vinegar. Children were asked to look at the containers' contents and tip them if they wished. They recorded their ideas on how the contents of pairs of containers were alike and how they were different. It was felt appropriate to limit the task in view of its repetitive nature. Therefore a set of comparisons with container A which held the piece of steel rod was requested. Children were free to record that there was no similarity (or even no difference) in their opinion. Samples in containers were used rather than materials from the display so that they would be standardised in quantity and also to avoid messiness.

This fourth activity was also designed to relate to the issue of children's conceptions of the states of matter. The materials were deliberately chosen as three solids having very different properties, and two liquids of a different nature. A steel bar is a hard, rigid, strong solid while cotton wool is a soft solid and the talcum power a powdery one. Treacle is a viscous liquid while vinegar is runny. It was therefore of interest to see whether the words 'solid' and 'liquid' would be used in making comparisons and whether properties characteristic of all solids or of all liquids would be referred to.

In contrast to this search for unsolicited use of the terms 'solid' and 'liquid', the fifth activity addressed the same issue directly. Children were asked to reveal their ideas about solids, liquids and gases by drawing pictures of them. These drawings were done side by side on a page divided into three sections and were annotated/labelled by the child or teacher. Children were free to leave a section blank where the word meant nothing to them.

2.3   Pre-intervention Interviews - Further Elicitation

Interviews in SPACE Project research have always been designed to extend the understanding of children's ideas emerging from the exploration experiences. That is, the interviews have been an attempt to probe and clarify the thinking that has emerged from the exploration experiences. In this research it was also necessary to extend coverage of issues not addressed directly in the classroom work. It had already been decided that some issues could be assessed more readily through an interview.

The interview schedule was not, however, drawn up until some feedback from the classroom activities had been received. This was obtained firstly through classroom visits when the progress of the activities could be seen and teachers could report what they had learned about the ideas of children in their classes. Secondly, some of the products from the activities, drawings, writing, etc. had been obtained and a preliminary scan was done to outline the thoughts they seemed to reveal.

The resultant interview schedule, which appears as Appendix IV, reflects the attempt to respond to the feedback received. Exploration experiences had elicited certain ideas which required further clarification. In addition, some aspects not touched on by these experiences required addressing. For example, it had not yet been ascertained whether children could make any links between properties of materials and their uses. There was, however, no possibility of piloting interview questions. The interviews themselves thus revealed other lines of enquiry and some questions proved more fruitful than others.

The interview was divided into sections with the following headings:

Section A	Meaning of hard and strong
Section B	Relating properties to uses
Section C	Concept of solid and liquid
Section D	Concept of gas
Section E	Changes on heating
Section F	Colourless liquid
Section G	Transforming metal - processes involved

The first section of the interview addressed children's understanding of some properties of materials (first issue). There were many properties that it would have been possible to ask children about. Since the exploration experiences had already covered the first issue from several angles, it was decided to limit enquiry to two properties, strength and hardness. These were chosen on the grounds that the adjectives 'strong, weak, hard and soft' had often been used by children during the exploration experiences. It was not clear, however, whether they were using the terms in a well-defined and separable way. Children were therefore asked how they would decide which of two materials was, firstly, harder and, secondly, stronger.

Section B of the interview addressed an issue not approached through the exploration experiences, that of relating properties to uses (third issue). During exploration, children frequently referred to how various materials were used. It was of interest to see whether they had ideas about why particular materials served a particular function. For each of four different materials, children were asked why they thought that particular material was good for making a particular object.

The next four sections of the interview all related to aspects of the states of matter - solid, liquid and gas (second issue). In the first of these, Section C, children were asked to classify some of the materials they had met previously. These were a steel rod, cotton wool, treacle, talcum powder and vinegar. Children were asked to say which they thought were solids and which liquids and to say how they had decided. They were also asked about any of the materials they had left unclassified.

Section D was intended to elicit ideas about gas. Children smelt some vinegar and were asked to talk about how they were able to smell it. They were also shown an 'empty' container and this was used to elicit their ideas about air.

Section E was designed to bring out some ideas about change of state. It is a line of enquiry which could be greatly extended but, in this instance, was not because previous SPACE research on energy (to be reported) and evaporation and condensation (reported) had already considered this theme. Children gave their opinions as to what would happen when each of three materials, a steel rod, cotton wool and vinegar, was heated to a very high temperature.

Finally, in this set of related sections of the interview, the purpose of Section F was to further examine the bounds of children's concept of liquid. Children were shown a container of colourless liquid and were asked what they thought it could be and how they could find out what it was.

Section E, 'Changes on heating', also addressed a further issue, that of change in general (fourth issue). On making a preliminary scan of children's ideas about the origins of materials (Activity 3 of the exploration experiences) it had become apparent that a limiting factor in considering where materials come from is the child's acceptance or rejection of the possibility that something can be changed. Thus, both Section E and Section G were designed with this in mind. While the former focused on changes brought about by heat, the latter was directed towards the possibility of changing the form of metal. Children were shown various forms of metal - powder, wire, wool and foil - and asked what they believed about the possibility of changing a metal rod into each of them.

The questions on the interview schedule indicated the areas in which there would be an attempt to elicit children's ideas. However, in order to maintain informality, interviewers were encouraged to rephrase the questions where children were unclear what was being asked. They also followed up children's lines of thought by asking additional questions in as spontaneous a manner as possible. The general guidelines for the interviews are given in Appendix V.

The interview schedule was extensive, covering the seven sections previously referred to. Allowing for a flexible response to clarify as well as probe children's ideas meant that interviews ranged from about 30 minutes to one hour. They generally lasted for about 45 minutes. This time is at the upper limit of that which is acceptable to the younger children in the sample and it might have been advantageous to have divided the interview into two sections carried out at different times. It was not possible to do this, however, because of the need to complete interviews within a short time period so that intervention work could start.

Where possible a sample of six children from each of the twelve classes was interviewed individually. Most teachers had been involved with the SPACE project for some time; since some had changed the age range taught, this meant a slight imbalance in the numbers teaching each age range. Thus, four taught infants, two lower juniors, one a mixed class of lower and upper juniors and five upper juniors. A larger sample had therefore to be taken from the lower junior classes and a smaller one from some upper junior classes to balance the numbers overall. The samples were balanced for gender and achievement and were randomly selected within these constraints.)

Members of the project team including the advisory teacher visited the schools and talked to children either in the classroom or in an otherwise unoccupied room. The interviews were conducted in an informal manner and every attempt was made to put children as much at ease as possible. In fact, in several classes children were not only used to expressing their ideas but also extremely keen to do so.

3. CHILDREN'S IDEAS

3.0   Introduction

This chapter describes the nature and frequency of occurrence of ideas prior to any attempts to help children to develop their thinking. All data are drawn from the participating group of schools; in some cases, children's ideas were elicited by means of relatively informal class-room techniques; alternatively, and this applies to most of the responses discussed in this chapter, a number of children representing a stratified sub-set of all those involved in the study were interviewed individually and their ideas recorded. Table 3.1 summarises the characteristics of the individually interviewed sub-sample.

Table 3.1   **Sample by age group, gender and achievement band (n=68)**

		Infants	Lower Juniors	Upper Juniors	
Low Achievers	girls	-	2	3	
	boys	4	3	5	17
Medium Achievers	girls	5	6	3	
	boys	6	5	4	29
High Achievers	girls	4	3	4	
	boys	4	4	3	22
30 girls 38 boys		23	23	22	

Achievement band was defined as overall scholastic achievement based on teachers' know-ledge of the particular children and general normative expectations of the year group as a whole. Although this is a fairly loose definition, it was relatively unambiguous to teachers and was operated with ease. It was emphasised that actual rather than potential performance should be the important criterion in allocating children to achievement bands 'high', 'medium' or 'low'.

The sample described in Table 3.1 can be seen to have achieved a reasonable balance in respect of some of the variables which might be assumed to have a significant impact on interview performance. (Children were selected randomly for inclusion in the interview sample, though those to whom the interview might have proved too stressful, in the teacher's judgement, would have been substituted. One such substitution was made when one child, a high achieving lower junior girl, showed apprehension about being interviewed. No other specific communication problems were encountered).

18

The pre-intervention sample described here is identical to that interviewed a second time (about eleven weeks later, following the classroom intervention and a school holiday) to establish any shifts in ideas subsequent to a range of classroom activities. Inevitably some children were lost to the sample due to illness, absence, etc. Table 3.1 records only those children for whom pre- and post-intervention data were available. Low-achieving infant girls apart, the interview sample includes a good representation in all cells, justifying cautious generalisations to the population as a whole.

Collection and analysis of qualitative data is an extremely time-consuming enterprise, potentially infinitely so as each extra interview question reveals further insights and invites further exploration. Boundaries must be defined by practical constraints. There was not the time available to pursue every issue raised in the pre-intervention interviews in the intervention, or through to the second round of interviews. This chapter reports the main ideas which children were found to be holding after a period of reflection but prior to a concerted effort on the part of their teachers to promote reconsideration. The next chapter, Chapter Four, describes the kinds of activities which teachers and children engaged in as part of that process of moving ideas along. Chapter Five then examines the shifts in thinking which are manifest between the two sets of interview data. To avoid repetition, this chapter illustrates the issues emerging from the 'starting' ideas by reference to material for which post-intervention interview data are not available: the issues are the same, the content may be different. Chapter Five presents some pre- and post-intervention data side by side for comparison. The issues dealt with in this chapter are those outlined in Chapter One and reflected schematically in Figure 1.2; within each major issue are some more or less discrete sub-sections, as follows.

3.1   Descriptions and Properties of Materials

    3.1.1   Classification of Materials
    3.1.2   Judgements about the Properties of Materials

3.2   Solids, Liquids and Gases

    3.2.1   Smelling Vinegar and Assumptions about Gases
    3.2.2   Identification of Materials as Solid or Liquid
    3.2.3   Identification of a Colourless Liquid

3.3   Uses of Materials

    3.3.1   Relating Properties of Materials to Uses

3.4   Origins, Manufacture and Changes in Materials

    3.4.1   Origins and Transformations of Materials
    3.4.2   Ideas about the Possibilities of Transforming Metal
    3.4.3   Predicted Changes on Heating

3.1 Description and Properties of Materials

3.1.1 Classification of Materials

One of the early issues which was explored with children was the question of how they attempted to classify objects in the world around them in terms of what those objects are made of. At a later stage in the programme, children's assumptions might be revealed through their actions on materials. Before this more active phase, it was of interest to address questions such as the following. Do children perceive similarities in the wide range of materials which make up individual objects? Are they aware of the nature of the materials of which the objects are made? What set names do they use when they attempt to classify materials? A classification exercise which entailed putting a range of everyday objects into sets was used both prior to and following a series of more actively investigatory classroom activities. The pre-intervention work took place in six schools in the Ormskirk area (see Appendix II for participating schools and teachers; this work was conducted during an earlier phase of the work than all other material discussed in this report). After a three week period during which some changes would have been likely to have been observed in the organic materials, children carried out a repeat classification. (Organic materials were disposed of before they became noxious.)

Figure 3.1   **Objects which children were required to put into sets 'according to what they are made of'**

steel wool	sand	plastic bag
apple	plastic cup	newspaper
nails	potato	white wood
brick	milk bottle top	wool
cotton cloth	copper pipe	rice
tomato	cutlery	pebbles
aluminium foil	white bread	candle
sugar paper	clay	lettuce
brown wood	brown bread	

Each of the six schools was provided with a set of materials and teachers asked children to put the items into groups according to what they were made of. The objects were available for observation and handling during the classification exercise. The method of recording varied from school to school. Some teachers asked children to write down their groupings. Younger children or those older children having difficulties with writing drew pictures which could be annotated during discussion with the teacher. Additionally, after three weeks, children were interviewed individually to establish their ideas about materials. During this three-week period, some of the organic materials would have decayed and were disposed of, though it was anticipated that the observed changes might have contributed to ideas about the nature of materials.

The experience of the individual interviews confirmed teachers' reports that children were willing to attempt the classification task as posed, and indeed showed evidence of some deep thinking about it. However, it became clear that many children were inclined to classify by reference to criteria other than those which the task had attempted to define, for example, by reference to what the item was used for rather than what it was made of. For instance, a six year old formed a set including the nail, bottle top, wire wool, copper pipe and wood. When asked why the wood had been included, the response was:

Because you could do something with it with the nail.

The wood had been linked to the nail by reference to its use; the set would otherwise have consisted entirely of objects made of metal. This child appeared to have a fairly clear idea about the variety of features common to metals - the objects included in the set of metals were quite disparate in appearance - yet seems to have been distracted by a strong association of usage which led to the inclusion of wood in the metals set. For other children, considerations of use or function completely overwhelmed any consideration of the material of which the objects were composed. Figure 3.2 illustrates a fairly typical range of set labels as used by a Year Five child.

Figure 3.2   Objects drawn in sets 'according to what they are made of'.

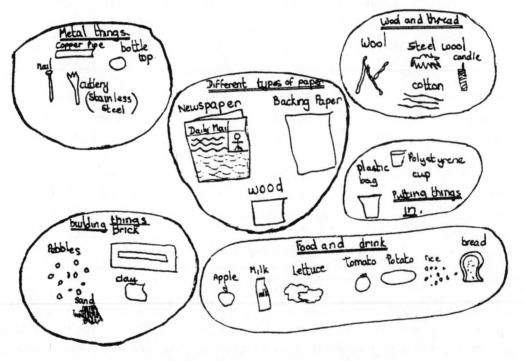

'Metal things' is an accurate compositional set; it lacks the steel wool, which has been placed in a set called 'Wool and thread'. The word 'wool' appears to have acted as a distraction and indeed, it was clear that many children did not regard the steel wool as a metal. Many children had difficulty knowing where to include the candle and in this case it has been placed alongside the cotton and wool. 'Building things', 'Food and drink' and 'Putting things in' are all functional categories. Newsprint, backing paper and the block of wood have been grouped together and labelled in a manner suggesting that the child is aware that paper is made from wood; this has led to the anomaly of wood being described by the set label, 'Different types of paper'.

A detailed analysis was undertaken of the responses of pupils from schools which had elicited children's ideas on an individual basis and had the most complete record of responses. This revealed over 50 different set names used to classify the range of objects in Figure 3.1. As suggested above, criteria other than those bearing on what the objects were made of (i.e. compositional) were used; the range of criteria were classified as follows.

1. *Compositional,* referring to what the objects were actually made of. Examples included: metal, plastic, wood, 'material', soil, stone, polystyrene, plant, etc.

2. *Functional* responses suggested what the objects could be used for and the sets were of objects which could be used in the same way: food, building materials, drink, things which hold things, to write on, art materials, everyday uses, things we use a lot, things we cannot eat, used in the kitchen, etc.

3. *Locational* properties described where objects might be found: natural/found in nature, inside things, outside things, found on the ground, seen on the beach.

4. *Perceptual* responses described how objects were perceived to have observable properties in common: hard, soft, shiney, solid/'unsolid', crackly, thin, heavy, squashy, shiney and smooth, changeable/not changeable, bendy, juicy, wet, silver, feel the same, make a noise.

5. *Manufactured,* man-made.

6. *Other Categories* was a catch-all group for the odd and idiosyncratic sets which children produced, often because they could not think where else to put odd remaining objects at the end of their classification: odd ones out, 'nothing' group, begin with letter 'p', etc.

Table 3.2   **Categories of response to classification task (per cent by age group)***

	Pre-intervention		
	Infants n=25	**Lower Juniors** n=57	**Upper Juniors** n=62
Compositional	57	53	38
Functional	-	35	53
Locational	-	8	1
Perceptual	43	3	8
Manufactured	-	1	-

*   These data are drawn from one infant school and two junior schools. N.B. individual children used more than one kind of category.

Table 3.2 suggests a higher incidence of references to perceptual properties amongst the infants, though this age group was not without an ability to comment on compositional properties. The juniors seemed to be more tempted to use a classification based on known functions of objects (perhaps as a result of their greater knowledge of how things are used), thus reducing the level of compositional responses. The locational category was less frequently used and the manufacturing aspect is recorded only to indicate its virtual absence from children's responses.

'Food' was classified as a functional rather than a compositional property; whether or not something is capable of being eaten seems to be a salient attribute of objects in children's perceptions. It is also probably fair to say that it is difficult for children to describe what organic objects are 'made of'. Attempts included 'vegetables', 'fruit', 'grain', 'health' and 'from trees and plants'. All of these are attempts to generalise.

The category 'metal' was of interest in that it is an accurate use of a generic term to describe a material and seems to be one of the earlier of such terms to be used spontaneously by children. As such, it might prove a useful example for teachers to use in clarifying the task, i.e. getting children to think about what things are made of rather than, for example, function. It would also be of interest to examine more closely which forms of metal are recognised as such, i.e., the critical features of 'metalness' in children's perceptions.

'Natural' tended to be used in the sense of not having been subjected to any manufacturing process, yet its converse, 'manufactured', was notable by its absence.

'Building materials' was a common functional category; it seems likely that one object such as the brick may trigger an association which engulfs other objects (sand and pebbles).

The use of the category 'material' is indicative of a specific vernacular usage of a word which is different from its scientific usage. 'Material' tended to be used to describe cotton or woollen fabrics; the generic term, in popular usage, tends to refer only to cloth.

3.1.2   Judgements about the properties of materials

It was the hope that the initial impressions resulting from the individual interviews, even before full and formal analysis was possible, would inform teachers and researchers about the kinds of activity which might be used profitably in the classroom as a way of extending children's thinking. It was assumed that an extension of the opportunities which they had already had to observe materials in an open and unstructured way would probably include classroom-based work involving further classification leading to judgements going beyond the more superficial properties of materials.

Children's understanding of two attributes of solids - hardness and strength - were explored during the initial interviews. From the array of objects which had been collected (which was, in most classrooms, on display close at hand), children were invited to choose two materials. They were then asked how they would decide which of the objects was the harder and which was the stronger. Two aspects of their responses were of particular interest; one point was the *process* aspect, the way in which they reached a decision, including the nature of the evidence to which they referred as sufficient to draw their conclusion; the second aspect was *conceptual*, the specific meaning attributed to the concepts 'hard' and 'strong'.

In considering the *process* aspect of children's decisions, a distinction was made between those responses which appeared to encompass an empirical test of some kind and those which appeared to be based on observational criteria alone. This distinction became a very fine line to draw on occasions; the criteria used are explicitly stated in the tables of data which follow in this section. Broadly, the responses judged to imply a test of some kind were of a more active, manipulative or invasive nature. Observations were less interventionist, involving looking, touching, or other more passive judgements. An example of a borderline response (and thus, though not commonly encountered, requiring a little more thought to categorise) would be, in relation to hardness, 'It hurts you if you bump into it'. Because of the implied passivity, this was grouped with the observational responses, as were those which suggested looking, or feeling. In contrast, a response which suggested scratching, cutting or filing the material to discover how hard it might be was treated as a test. For the majority of responses, this categorisation was unproblematic. The first two responses quoted here were treated as observations, the third as an implied test.

The brick [is harder]. Because it will drop on some people's toe
and hurt you.   Y1 B L

The bigger one would be harder.   Y1 B H

Try dropping them on the floor and see which breaks. If it broke,
it wouldn't be hard. You can't really break it without something
to help you.   Y3 G H

Although the division between the two qualities of response was not a sharp boundary, the analysis was felt to be worth attempting because of the importance within the project rationale of process skills supporting conceptual understanding. Children would be encouraged, during the intervention period, to look for evidence and investigate their ideas rather than make superficial observations and assertions. The extent to which an inclination to seek empirical evidence was already in place was consequently of particular interest to the research (as would be any shifts in this area, following intervention).

The criteria which children expressed as the evidence which they would seek to determine which of two chosen materials would be judged to be harder are shown in Table 3.3.

---

1   A code is used to identify the characteristics of members of the individually interviewed sample as follows:

R	Reception	B	Boy	H	High Achiever
Y1	5 - 6 years	G	Girl	M	Medium Achiever
Y2	6 - 7 years			L	Low Achiever
Y3	7 - 8 years				
Y4	8 - 9 years				
Y5	9 - 10 years				
Y6	10 - 11 years				

Table 3.3   Criteria used to judge 'harder'

	Pre-intervention		
	Infants **n=23**	**Lower Juniors** **n=23**	**Upper Juniors** **n=22**
Observational Criteria			
Size, thickness	4 (1)	–	–
Heaviness	22 (5)	4 (1)	18 (4)
Appearance	9 (2)	–	5 (1)
Feel, touch, texture (rough, smooth, 'it hurts you')	13 (3)	9 (2)	5 (1)
'What it's made of'	4 (1)	9 (2)	14 (3)
Criteria Implying Test			
Makes a sound on impact	–	4 (1)	–
Impresses, is malleable (dents, or can be squeezed)	4 (1)	9 (2)	32 (7)
Resists intrusion, cutting abrasion (e.g. use of nail, saw or file)	13 (3)	13 (3)	5 (1)
Has structural rigidity	22 (5)	4 (1)	9 (2)
Breaks/shatters on impact or when dropped	–	39 (9)	5 (1)
Bends under load	–	–	–
Other	–	9 (2)	5 (1)
Don't know	9 (2)	–	5 (1)

The 'heaviness' of an object was the most frequently cited observational feature used to judge hardness; whether children were intending to refer to mass or density is not clear, but in either case, the criterion of heaviness is not helpful simply because it is inaccurate. A number of children referred to the potential of hard objects to cause them bodily hurt, and it is possible that the use of the term 'heavy' is used with similar connotations. The next most common category was the feel or texture of the material (nine per cent of the sample), with the remaining observational responses mentioning what the object was made of or its appearance - 'It looks hard'.

Thirty-eight per cent of the sample overall made observational responses (52 per cent infants, 22 per cent lower juniors, 41 per cent upper juniors).

Of the responses which seemed to imply a test, one of the most frequently mentioned indicators of hardness was whether or not the material would take an impression; some children expressed this as a capability of being squeezed or dented. About one-third of the upper juniors offered this kind of response.

The most frequently mentioned indicator of hardness mentioned by the lower juniors - 39 per cent of this age group - was whether the material would break or shatter on impact, or when dropped.

> *Get something like a hammer and try to break it and which ever*
> *you couldn't break would be the hardest one.*　　　　　　　Y3 G H

Amongst the infants, yet another response was the most common: whether or not the material could be stacked or built up, i.e. whether it had structural rigidity.

> *The brick is hardest because we build houses with it and houses*
> *don't fall down.*　　　　　　　Y6 G L

Fifty-seven per cent of the interview sample used a criterion for hardness which implied a test, (39 per cent of infants, 78 per cent lower juniors and 55 per cent upper juniors). The ratios of observational to 'test' responses were thus 52:39 for infants (two 'don't know' responses), 22:78 for lower juniors and 41:55 upper juniors (one 'don't know').

A similar analysis, summarised in Table 3.4, was completed for children's comparisons of two chosen objects to decide which might be 'stronger'. Responses were categorised using the same criteria as for 'harder'.

Table 3.4   Criteria used to judge 'stronger'

	Pre-intervention		
	Infants n=23	**Lower Juniors** n=23	**Upper Juniors** n=22
Observational Criteria			
Size, thickness	4 (1)	–	–
Heaviness	35 (8)	13 (3)	5 (1)
Appearance	4 (1)	4 (1)	–
Feel, touch, texture (includes smooth, rough, 'it hurts you')	13 (3)	4 (1)	–
'What it's made of'	9 (2)	–	5 (1)
Criteria Implying Test			
Impresses, is malleable (dents, or can be squeezed)	–	13 (3)	–
Resists intrusion, cutting abrasion (e.g. nail, saw, or file)	9 (2)	30 (7)	5 (1)
Has structural rigidity	9 (2)	4 (1)	23 (5)
Breaks/shatters on impact or when dropped	9 (2)	26 (6)	41 (9)
Other	–	4 (1)	9 (2)
Don't know	9 (2)	–	14 (3)

The overall level of observational responses was slightly higher (38 per cent for strength compared with 32 per cent for hardness). While the proportion remained constant for the lower juniors, the infants offered more, the upper juniors fewer observational criteria than had been the case for judgements about hardness.

The ratios of observational to 'test' responses for 'stronger' were 65:26 for infants (with two 'don't knows'), 22:78 for lower juniors and 9:77 for upper juniors (with three 'don't knows').

The data give the impression that the distinction between the terms 'hard' and 'strong' is not at all clear to the majority of children. There is some indication that hardness tended to be associated with malleability while strength was related more often to the capacity to survive impact. This idea might be closer to a notion of the property of brittleness than the property of strength; a material's strength has more to do with its capacity to avoid breaking when subjected to tension or compression. No children offered a response referring to a material's capacity to avoid breaking when subjected to loads as an indication of strength, though there were more responses drawing on the idea of structural rigidity than had been apparent in relation to hardness.

It became clear that the particular materials under consideration influenced the operational definitions which children used to define the properties 'hardness' and 'strength'. This point is discussed further in section 5.1.2.

Children had been invited to comment on two properties and their responses had been subjected to similar analyses concerning the use of observation or an implied test in accumulating information upon which to base their decisions. It was of interest to determine the consistency with which they used particular kinds of reasoning to support their judgements. This information is summarised in Table 3.5.

Table 3.5   **Nature of criteria used to judge 'hard' and 'strong' (children's own selection of materials)**

| | Pre-intervention | | |
	Infants n=23	Lower Juniors n=23	Upper Juniors n=22
Reference to observation of both hardness and strength	44 (10)	4 (1)	5 (1)
Reference to test of both hardness and strength	26 (6)	61 (14)	32 (7)
Inconsistent use of test and observation	13 (3)	35 (8)	46 (10)
Other combinations (includes 'don't know', etc.)	17 (4)	– 	18 (4)

More than two-fifths of the infants were consistent in their use of observational criteria to support their judgements; only one each of lower and upper juniors showed similar observational consistency. One-quarter of the infants and one-third of the upper juniors were consistent testers; the lower juniors out-performed both of these other age groups with 61 per cent being consistent testers.

Thirty-nine per cent of infants suggested a test on at least one of the two occasions; the equivalent figure for the lower juniors was 96 per cent and for upper juniors, 78 per cent.

3.2   Solids, Liquids and Gases

3.2.1   Smelling vinegar and assumptions about gases

The sample of brown coloured malt vinegar was used as the starting point for the next area of questioning. The lid of the sample container was removed and the container was gently moved to and fro at a distance of a metre or less from the child until it was clear that the vinegar had been smelt. The interviewer then probed the child's understanding of how it was that the vinegar had been smelt, using questions such as 'How are you able to smell the vinegar?', and 'What happens to let you smell the vinegar?', with the intention of probing whether children had any notion of a movement of vapour from the vinegar to their noses. The interviewers were conscious of the fact that the wording of the initial question and follow-up probes was extremely important. The intention was to go beyond the simple, everyday response that children were likely to give, describing smelling as a self-evident phenomenon requiring no further explanation; in indicating that some further consideration was required, leading questioning had to be avoided.

In everyday usage, smelling might sometimes be regarded as a passive activity while at other times it might be viewed as something active; sometimes smells are invasive and unavoidable, while at other times the perfume of a bloom or the source of a gas leak might be actively sought by sniffing samples of air. This ambiguity of passive and active elements in the act of smelling is problematic when the intention is to gain information about children's understanding of the mechanism involved. The sample of vinegar could not be described as producing an overwhelmingly invasive smell in the context of the interviews. A purely active interpretation of smelling might be construed in some instances as being ego-centric; a purely passive description in which the smell is entirely attributed to the source is limited, but at least acknowledges the role of the material world in the act of smelling. A more scientific explanation would need to refer to the interaction between the material and the nose. This would eventually become elaborated into a materialist particle model in which it is understood that free floating particles of the smelt material produce characteristic sensory signals in the organ of smell. An intermediate understanding might attribute a role to the air with which the molecules of the smelt material mixes, the air being capable of wafting the smell towards someone.

Table 3.6 summarises children's responses to the question of how vinegar was smelt.

Table 3.6   Explanations of how vinegar was smelt

	Pre-intervention		
	Infants n=23	**Lower Juniors** n=23	**Upper Juniors** n=22
Reference to role of nose only (e.g. by sniffing, by your nose)	26 (6)	26 (6)	18 (4)
Reference to property of vinegar only (e.g. 'It's strong')	13 (3)	26 (6)	23 (5)
Reference to smell/fumes, etc., plus role of nose	35 (8)	35 (8)	18 (4)
'Smell' plus 'nose' plus interpretation (brain/head)	–	4 (1)	5 (1)
Other	9 (2)	–	14 (3)
Reference to 'air' as having a role in smelling	13 (3)	17 (4)	36 (8)

31

About a quarter of the sample at all ages (though slightly fewer upper juniors) explained smelling by reference to the action of the nose only. It must be assumed that this explanation was regarded by this group as sufficient, since it persisted despite probing during the interviews.

Q.   How do you smell [the vinegar]?
R.   *With your nose.*
Q.   How are you able to smell with your nose when it's so far from the vinegar?
R.   *Cos your nose has got a gap on two sides.*
Q.   What do they do?
R.   *Smell things.*
Q.   How do they smell things?
R.   *Because if you didn't have a nose you couldn't smell anything and if you did have a nose, you could.*   Y1 B M

Q.   How do we smell it when we're some way from it?
R.   *Because we're clever. We smell things with our noses.*   Y1 G M

R.   *Well, you can't smell it when the top's on and you can't smell it when you're far away.*
Q.   Why not?
R.   *Because your nose isn't that big.*
Q.   So how do you smell it?
R.   *With your nose.*   Y1 B H

Again, about one-quarter of the sample of upper and lower juniors (and half as many infants) referred only to a property of the vinegar as the factor responsible for the smelling sensation:

    It's strong, so I can smell it at a distance.   Y5 B L

    Because it has a strong scent; I don't know what happens.   Y5 G L

About one-third of infants and lower juniors and about half as many upper juniors used an explanation referring to an interaction between the object smelt and the organ of smell, though these responses did not imply use of a particle model. An elaboration of this basic interaction model which was in evidence indicated awareness of a psychological dimension to the process of smelling - the interpretation of the sensory information by the brain. This kind of response was offered by just two children:

    Well, the smell goes up your nose and it sort of goes to your head and tells you what it is. It's a liquid and it has a strong smell on it ... and you can tell whether it's a liquid or a powder. You can also tell if it's got a strong smell.   Y3 G H

    In your nose you have things that you remember it in your brain.   Y6 B M

Another aspect which was noted was the incidence of reference to the air as having a role in the process of smelling; this increased with age to about one-third of upper juniors making direct reference to air.

> *It's sweet and there's a lot of smell, so you can smell it. The smell is coming up to you. Instead of trying to breathe, a gentle wind blows it over. It blows it if you are on the right side, and it goes up your nose.*　　Y3 B M

> *The smell falls out into the air and with the air moving it comes towards your nose. If you move your head, it will move. The smell's been trapped and rose to the top of the lid and when you take the lid off, it comes out.*　　Y5 G M

Establishing what precisely children assumed might be smelt in the activity of smelling was difficult, primarily because of the limitations of vocabulary associated with a rather limited sense modality (compared, for example, with the vocabulary of vision or hearing). Consequently, what was smelt by most children was the 'smell', a tautology which is perfectly adequate for conveying meaning in everyday life (see Table 3.7) or equally economical, the 'vinegar'.

Table 3.7　　Reference used to describe what was smelt

	Pre-intervention		
	Infants n=23	Lower Juniors n=23	Upper Juniors n=22
Vinegar	9 (2)	4 (1)	27 (6)
'Smell'	30 (7)	52 (12)	59 (13)
'Fumes'/'smoke'	4 (1)	4 (1)	–
No reference to object, only action of smelling	35 (8)	26 (6)	14 (3)
Other	8 (2)	4 (1)	–

About a third of infants, half the lower juniors and nearly two-thirds of the upper juniors described what they had smelt as the 'smell'; the majority response amongst the infants was to refer to the act of smelling rather than the object. This latter response was encountered with decreasing frequency in the two older age groups, perhaps lending support to the suggestion that it is an ego-centric response, more likely to be encountered amongst younger children.

> *I can smell it because it's in the bottle and I can easily smell it.*   Y2 B L

Two children referred to 'fumes' or 'smoke'.

> *You're sniffing up the fumes out of it. I'm not exactly sure, but if you open it, the fumes might come out. They spread around, sort of like germs do.*   Y3 B H

> *When you sniff, it gets to your nose and you can feel it. The smell is like a smoke that comes up to your nose and you can smell it.*   Y2 G H

The word 'fumes', if linked to a perceptible (and safe) experience, might be a useful mediating word between the concrete starting point (vinegar, or whatever) and the abstract notions of particle or gas. 'Fumes' have a more tangible, or at least perceptible, quality than the abstract terms. Indeed, the use of the word 'fumes' in the context of smell is reminiscent of the use of 'mist', 'spray', etc. in children's attempts to construct a perceptible bridge in their understanding of the equally abstract evaporation process.

Assumptions about gases

An empty container, similar to that used to contain the vinegar, was the physical stimulus used to initiate an exploration of children's ideas about the presence and nature of air. Children were shown the empty container and asked whether they thought there was anything in it. Almost without exception, the response was that the container was empty; specific results to this question are not reported because it was clearly insufficiently contextualised for children to appreciate that the interviewers' focus of interest was air. The next step was to offer a counter-suggestion in the form, 'Somebody else said that the container is not empty. What do you think? What were they thinking of?' The wording was adapted to individual children but shared a common form in not specifying any characteristics of the possible contents of the container. The context of the question was thus, implicitly, 'Do you think there is anything in the container having no apparent perceptible qualities?' Table 3.8 presents a summary of children's responses to this elaborated form of the question.

34

Table 3.8 **Responses to the counter suggestion that the container might not be empty**

	Pre-intervention		
	Infants n=23	Lower Juniors n=23	Upper Juniors n=22
Suggested that the 'empty' container contained air	22 (5)	96 (22)	73 (16)
Persisted with idea that container was empty	74 (17)	5 (1)	23 (5)
Other response	–	–	5 (1)
Don't know	4 (1)	–	–

Less than a quarter of the infants suggested that there might be air in the container following the counter-suggestion, while almost all the lower juniors and about three-quarters of the upper juniors offered a similar view. It would appear that, given a sufficient contextual cue, there is a fairly widespread awareness amongst the 7-11 age range of an invisible, odourless and intangible material - the air - being present in the world around them. On the other hand, children's first thoughts may not take the air into account.

Once the idea of the presence of air had emerged, many children volunteered further information about its nature and properties. Otherwise, the ideas held about the properties of air by those children who indicated awareness of its existence following the counter-suggestion were established by further indirect questioning of the kind, 'What do you know about air? What's it like?' The frequencies with which a range of properties were mentioned by the three age groups are presented in Table 3.9.

Table 3.9   **References to properties of air**

	Pre-intervention		
	Infants n=23	**Lower Juniors** n=23	**Upper Juniors** n=22
Invisible/you can't see it	17 (4)	30 (7)	45 (10)
Odourlessness	–	4 (1)	5 (1)
Associated with felt movement, wind, breeze, etc.	17 (4)	22 (5)	14 (3)
Reference to associated temperature sensation	4 (1)	13 (3)	–
Life sustaining/for breathing	4 (1)	48 (11)	41 (9)
Mean number of above properties referred to	0.4	1.2	1.0

The properties of air reported in Table 3.9 are those spontaneously suggested by children. An increasing awareness with age of air existing around them, despite its invisibility, is apparent; however, relating the data in Table 3.9 to those in Table 3.8, it becomes apparent that four of the five (80 per cent) infants acknowledging the presence of air referred to its property of invisibility. The equivalent percentages for the upper and lower juniors were 32 per cent and 63 per cent respectively.

> *It might be air, because you can't see it.*   Y1 G M

> *You can't feel it. You can't even see it. It's all around you.*   Y1 B H

> *The air, if you close a lid, it's trapped in the bottle. If there's nothing in it, it's a solid shape cos air's in it. But if you take the lid off the air floats out. You can't see the air, but sometimes you think there is nothing in there cos you can't see anything, but it isn't a matter of seeing, it's a matter of knowing.*
>   Y3 G H

Only two children, one lower junior and one upper junior, referred to odour. The former asserted that air was not a gas because it could not be smelt, while the latter suggested that 'it doesn't smell of anything'.

Felt movement of air, perhaps because it was a tangible quality which made air more concretely accessible conceptually, was mentioned relatively frequently.

> *You can't see it. It's like little breezes coming on you ... It cools you down*
> *a bit. It can do a lot of powerful things to people's houses.*   Y3 B H

The response above was no doubt influenced by storms which the child had experienced recently; it also contains a reference to temperature sensation associated with air - once again, a tangible effect.

> *It's got fresh air in it. It breezes all your hair back. Sometimes it makes*
> *you cold.*   Y3 B M

The mean numbers of properties of air mentioned by the sample as a whole are shown in Table 3.9, and suggest that the infants are less productive than the juniors in this respect. However, looking at the infants who expressed awareness of air, as a sub-group, they were more productive of suggestions about properties of air than their older peers, on average. The modal response of the 'air aware' infants was to indicate one property; the modal response of the lower juniors was one property, that of the upper juniors, two properties.

Table 3.10   Ideas about the extent of air

	Pre-intervention		
	Infants **n=20**	**Lower Juniors** **n=22**	**Upper Juniors** **n=22**
Everywhere/all over/ all around	30 (6)	68 (15)	77 (17)
Extent limited locally, e.g. playground	–	23 (5)	–
Don't know	10 (2)	–	–
No response	60 (12)	9 (2)	23 (5)

Children's ideas about the extent of air are shown in Table 3.10. All the infants who offered a response suggested that air was to be found everywhere, or all around.

> *Air is colourless and all around people and people breathe air.*   Y2 G H

> *Air is something you can't feel and it's everywhere.*   Y2 B M

No children indicated an awareness of the extent of air being limited to the Earth's atmosphere. Five lower juniors suggested that air has a localised presence of a much more immediate nature.

Q.   Where do we find air?
R.   *Over there, in the room [points].*
Q.   Is there air outside?
R.   *Only in the playground.*   Y3 B L

3.2.2   *Identification of materials as solid or liquid*

Five prepared materials were shown to children. Each was contained in a sealed glass container to enable it to be observed and safely handled. The materials were a length of steel rod, vinegar, brown treacle, cotton wool and talcum powder. Children were asked to decide whether each material was a solid or a liquid; the choice was not forced or seen by children to be exhaustive. Many responded that they did not know, or that some of the materials were neither solid nor liquid. The same task using identical materials was repeated post-intervention and to avoid repetition, details of responses and comparisons of results are presented there, with a brief summary only at this point.

Broadly speaking, the pre-intervention responses indicated that the metal rod and the vinegar were relatively unproblematic to classify as solid and liquid respectively. The viscosity of the treacle seemed to challenge the core attribute of 'runniness' which most children wanted to see in a liquid. The cotton wool was not an archetypically strong, hard solid so its softness and malleability posed a challenge. The powder shared a liquid's capability of being poured, and lacked some solids' hardness and structural integrity.

Having asked children to classify the small range of given materials described above, a complementary classroom activity was designed to elicit their own suggestions of materials which could be included within each of the three sets, 'solid', 'liquid' and 'gas'. Teachers asked children to divide a page into three sections and to draw examples of solids, liquids and gases. Some of the examples children selected from the table of classroom materials, but it was made clear that they could include whatever additional materials they wished.

The same drawing task was repeated following the classroom activity period. The drawings of the interview sub-sample were scrutinised and some recurring elements in children's responses were noted. A mark scheme was prepared which enabled the incidence of certain salient features of response to be quantified. The analyses are reported in detail in section 5.2.2. while some more general points are illustrated at this juncture.

Figure 3.3 exemplifies some recurrent features as well as some which were seen more rarely. Firstly, it is unusual in that there are equal numbers of solids, liquids and gases shown - three of each. More commonly, there were far more solids depicted than liquids, and more liquids than gases. Secondly, it is unusual in illustrating a named gas - oxygen.

38

A recurring characteristic of the solids which children chose to draw was that they tended to be of the strong and hard variety. In the process of their mental searches for liquids and gases, children often seemed to be spurred by word associations or word 'triggers'. For example, Figure 3.3 includes an example of washing up 'liquid', and a 'gas' stove. It is probable that the otherwise unlikely inclusion of a gold object was prompted by the cliché, 'solid gold'. The other two solid objects, the metal rod and the brick wall, are emphatically in the centre of the class of objects which children tend to see as unambiguously solid.

Figure 3.3   Solids, liquids and gases: Drawing 1

The named gas was an unusual feature, but the other two gases illustrate some recurring features of the gases which children selected. The aerosol can is interesting in that what children are able to see coming out of it is a perceptible spray; one of the great difficulties of becoming familiar with even the most common of gases is their odourless invisibility. The spray canister overcomes both of these difficulties (even though what is seen is not actually a gas, but droplets of liquid), with the additional perceptible quality of a hissing noise as the spray and gas mixture escapes.

The third example in Figure 3.3, the gas stove, is an example of a word-'triggered' response. It also has another feature which was common, the association of gas with the function of heating or the property of combustion.

Figure 3.4 includes more examples of materials representing each state than the average, but the proportion of each - 10 solids, 7 liquids and 3 gases - is fairly typical. The ten solid objects are, once again, all strong, hard and structurally rigid with metal, wood, stone and plastic predominating.

Five of the seven liquids are drinkable, two not; there tended to be a strong association between the word 'liquid' and the notion of drink.

Figure 3.4   Solids, liquids and gases: Drawing 2

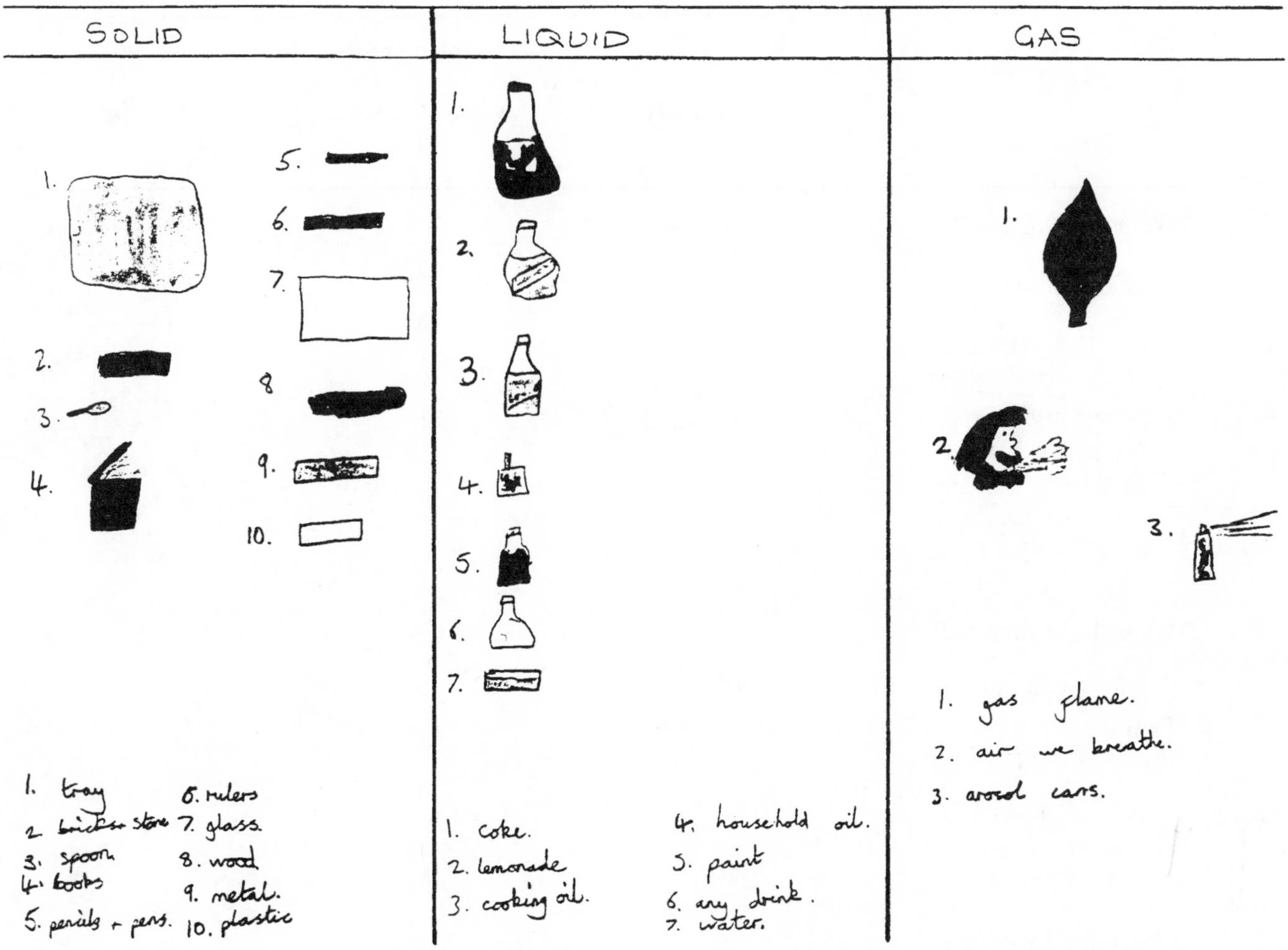

SOLID	LIQUID	GAS

Solid:
1. tray
2. bricks + stone
3. spoon
4. books
5. pencils + pens
6. rulers
7. glass
8. wood
9. metal.
10. plastic

Liquid:
1. coke.
2. lemonade
3. cooking oil.
4. household oil.
5. paint
6. any drink.
7. water.

Gas:
1. gas flame.
2. air we breathe.
3. aerosol cans.

The drawings of gases illustrate three points having general applicability. Once again, in the drawing of the gas flame, the word-trigger and the association with combustion is seen. The 'air we breathe' was also frequently referred to by older children; it seems likely that children have encountered the technical use of the term 'gas' in the context of studies of the process of breathing. The 'aerosol can' appears again in Figure 3.4, the perceptible quality of the emission clearly depicted in the drawing.

3.2.3 Identification of a colourless liquid

Children were shown a small quantity of water in a sealed transparent container and were asked, 'What could this be?' Almost all identified the liquid as water; one infant was unable to offer any suggestion; three lower juniors and one upper junior mentioned the possibility that the liquid might be other than water. A mild counter-suggestion was then offered in the form, 'What else could it be?' Responses to the initial question and to the counter-suggestion are summarised in Table 3.11.

Table 3.11   Identification of colourless liquid

	Pre-Intervention		
	Infants n=23	**Lower Juniors** n=21	**Upper Juniors** n=22
'What could this be?'			
Liquid identified as water	96 (22)	86 (18)	95 (21)
Suggestion that liquid might be other than water	–	14 (3)	5 (1)
Don't know	4 (1)	–	–
'What else could it be?'			
No acknowledgement of any possibility other than water	78 (18)	14 (3)	27 (6)
A beverage other than water (vodka, mineral water, lemonade, etc.)	13 (3)	67 (14)	46 (10)
A non-drinkable colourless liquid (turpentine spirit, acid, sea-water, etc.)	–	14 (3)	23 (5)
Other	–	5 (1)	–
No response, don't know	9 (2)	–	5 (1)

Three-quarters of the infants admitted of no other possibility than that the liquid was water, compared with a fifth of the juniors; 13 per cent of the infants were willing to entertain the alternative, but all opted for the possibility of another drink of some kind.

The patterns of response of the lower and upper juniors, though not the frequencies, were similar. Most thought the liquid might be a beverage of some kind (about two-thirds of the lower juniors, just under half the upper juniors). Suggestions included: lemonade, flat lemonade, '7 Up', 'Perrier', vodka, white wine, tonic, gin and medicine. A minority (14 per cent of the lower juniors, 23 per cent of the upper juniors) suggested that the colourless liquid might be a non-drinkable substance: sea-water, acid, vinegar, turps (turpentine spirit).

Having established the child's ideas about what the liquid might be, the interviewer then posed the problem of how children would find out whether or not the colourless liquid was, in fact, water. Table 3.12 summarises responses to this question; children often suggested more than one test or observation.

Table 3.12   Evidence for the identification of a colourless liquid as water

	Pre-intervention		
	Infants **n=23**	**Lower Juniors** **n=21**	**Upper Juniors** **n=22**
Smell only (without tasting)	13 (3)	9 (2)	18 (4)
Taste only (without smelling), drink	35 (8)	14 (3)	36 (8)
Smell and taste	–	9 (2)	23 (5)
Shake/look for bubbles	78 (18)	14 (3)	27 (6)
Look for transparency ('see through', etc.)	22 (5)	14 (3)	5 (1)
Feel for 'wetness', put finger in	9 (2)	5 (1)	5 (1)
Immerse something	–	5 (1)	–
Recognition of need for caution explicitly mentioned	9 (2)	–	5 (1)

As so often with young children, expressions of how a hypothesis would be tested tended to be in terms of how a belief would be confirmed. Thus, with the majority of the infants assuming that the colourless liquid was a drink, the tests suggested were to decide whether it was water or a carbonated drink. Just over three-quarters of the infants suggested shaking the container to look for bubbles; the presence of bubbles, it was assumed, would determine the liquid to be lemonade or a similar drink.

> *Lemonade's got bubbles in it, water hasn't.*   Y4 B L

About one-fifth of the infants affirmed transparency as a determining factor in deciding that the liquid was water, this property being less frequently cited with increasing age. Six per cent of the sample suggested that they would feel for 'wetness', which they seemed to regard as a unique property of water. (To some extent it is probably true to say that water has a characteristic 'feel' to which children are well attuned. This feel would include intuitive notions about viscosity and evaporation rate, perhaps, but the notion of immersing the fingers in an unknown colourless liquid is obviously one to be discouraged.)

Q.   Could it be anything else that was not water?
R.   *No.*
Q.   Suppose somebody said to you that it was not water. How would you find out?
R.   *Well, I would find out by putting my fingers in it. Your fingers are to feel and you know that it is water because your fingers are wet.*   Y3 G H

> *You could tell by looking. Lemonade looks brighter. You could touch it. If you put your finger in lemonade, all bubbles comes round your finger.*   Y3 B L

Not surprisingly, given children's assumptions that the liquid was most likely a drink, and incidentally emphasising the general lack of awareness of safety issues, smelling and tasting figured prominently in children's proposed search for evidence as to the identity of the liquid.

> *Take the top off and smell it. If it didn't smell, it would be water and if you tasted it, it could be lemonade or Perrier water.*   Y3 G M

> *Drink it. Water isn't fizzy, lemonade is.*   Y3 B M

Fourteen per cent of the sample said they would smell the liquid; 29 per cent suggested tasting without prior smelling; 11 per cent of the whole sample, all juniors, advocated smelling and tasting. This response was not necessarily symptomatic of an implicit caution. In total, 53 per cent of the sample suggested smelling and/or tasting the colourless liquid to determine whether it might be water. Two infants and one upper junior made unsolicited comments about the need for caution.

This section of the discussion was drawn to a close by the interviewer emphatically stressing the danger of smelling or tasting unknown liquids.

3.3   *Uses of Materials*

3.3.1   *Relating properties of materials to uses*

Children were asked the following questions:

> *Why do you think wood is a good material for making furniture?*
>
> *Why do you think rubber is a good material for making tyres?*
>
> *Why do you think wool is a good material for making clothes?*
>
> *Why do you think metal is a good material for making coins?*

In each case, responses to the initial question were probed to see whether the child could elaborate, or refer to more than one property of the material. Exceptionally, these questions were posed in the absence of specific physical examples of the items under discussion, though, of course, in the case of clothes and furniture, there were examples in view. Some examples of the other materials were likely to have been included in each class's display of materials, though not in the context of their uses. The assumption was made that all the objects were sufficiently familiar to children not to pose an unreasonable demand on memory.

It was found helpful to classify children's responses using the following general criteria:

Functional; referring to attributes which suited the material to its use. Responses might include strength and durability in relation to wood, the strength and elasticity of rubber, insulating properties of wool, and strength and durability of metal.

Manufacturing; referring to the 'workability' of the material such as the ease of cutting and fixing of wood, the weaving or knitting of wool, the moulding of rubber and metal.

Aesthetic; referring to the pleasing attributes, particularly of appearance, of the material. In most cases, classification using this criterion was clear-cut, but it did occasionally blur into functional properties. Children referred to the possibility of painting, staining and polishing wood, the quiet ride provided by rubber tyres, the softness and potential for dyeing wool and the shiney appearance of coins.

Economic; referring to the abundance, low cost of obtaining or manufacturing the material, or the renewable nature of the resource.

The data in Table 3.13 refer to the patterns of response in relation to wood and rubber only, but a similar outcome was in evidence in relation to suggested properties of wool and metal also. (Section 5.3.1 describes children's ideas about properties and uses of wood and metal as they emerged following the period of classroom intervention activities.)

Table 3.13   Mean number of references to functional, manufacturing, aesthetic and economic properties of wood and rubber.

	Pre-intervention		
	Infants n=23	**Lower Juniors** n=23	**Upper Juniors** n=22

Mean number of suggestions:

Wood for Furniture

Functional	0.7	0.8	1.1
Manufacturing	0.2	0.4	0.4
Aesthetic	0.1	0.1	0.4
Economic	0	0.1	0.1

Mean number of suggestions:

Rubber for Tyres

Functional	0.7	1.0	1.8
Manufacturing	0.2	0.1	0.1
Aesthetic	0.1	0	0
Economic	0	0	0.1

Functional properties were by far the most common kind of references which children produced. The sources of their knowledge can only be speculated upon but it seems likely that they were, in many instances, calling on knowledge and experience from beyond the school context.

> *Wood is good 'cos it's strong. It can hold things up for a long time.....If it's out in the rain it won't rot for a long time. Wood used to be trees and trees never rotted when rain came on them.*
> (Wood for furniture, functional property)   Y6 B M

Manufacturing qualities, usually expressed as the relative ease with which a material might be worked, were the next most frequent kind of property to which children made reference.

> *Wood is good because it won't collapse apart. You nail it together and it sticks together. Wood is smoother than brick because its not as strong.*
> (Wood for furniture, manufacturing property).   Y2 B H

*Well, it's easy to make...and it doesn't get holes in it very often and it's
very strong.*
(Rubber for tyres, manufacturing and functional properties).   Y3 G H

Aesthetic properties were mentioned fairly infrequently, though perhaps understandably less
frequently in relation to rubber than, e.g., wool.

Because it's soft, and it's nice to wear.
(Wool for clothes, aesthetic property)   Y5 G L

Wood looks nice if it's stained.
(Wood for furniture, aesthetic property)   Y5 B H

Economic considerations were voiced very infrequently and their inclusion in the analysis is
to underline the significance of their absence.

*Well, it's stronger really . . . and I suppose it's the only thing about,
because there are loads of trees which you can get wood off.*
(Wood for furniture, functional and economic properties)   Y3 G M

It's warm. Sheep can grow more wool for us again.
(Wool for clothes, functional and economic properties)   Y6 G H

To some extent, responses were context-specific; younger children were able to generate
more suggestions about the utility of wool for clothes than of metal for coins, for example.
Table 3.14 summarises the mean number of suggestions by age group for each of the four
materials and products discussed.

**Table 3.14   Overall mean number of properties of wood, rubber, wool and metal
related to uses**

	Pre-intervention		
	Infants n=23	**Lower Juniors** n=23	**Upper Juniors** n=22
Wood for furniture	0.9	1.3	2.0
Rubber for tyres	1.0	1.1	1.9
Wool for clothes	1.1	1.5	2.0
Metal for coins	0.7	1.2	1.7

Very roughly, the infants generated about one suggestion per material, the upper juniors
twice that number, with the lower juniors somewhere between the other two groups.

3.4 Origins, Manufacture and Changes in Materials

3.4.1 Origins and transformations of materials

Several variations on the theme of using drawings to establish children's ideas in explicit form have been used as a research technique within the SPACE Project. In an attempt to unravel children's ideas about the origins and transformations of materials, they were asked to take a product back through time, drawing any states or transformations which they were aware of or considered to be important. The picture strip technique had been used effectively in other instances, for example, to show a sequence of events in time, such as rusting. It was anticipated that a similar exercise, but going backwards through time, might pose a greater challenge, but in the event, most children managed the technique without insurmountable difficulties.

Two manufactured materials were used as starting points: a small piece of coloured cotton fabric and a metal spoon. Teachers introduced the task, demonstrating what was required by tracing back the origins and transformations of another material, a plank of wood, to illustrate the drawing technique to the whole class. An example of the technique, referring in this case to the origins of flour, is shown in Figure 3.5.

Figure 3.5   Drawing the origins of materials

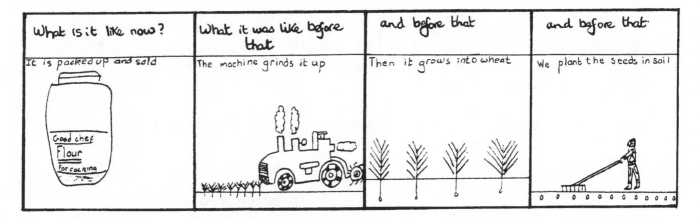

Children were then asked to make their own drawings to trace the origins of the cotton fabric and metal spoon. The contents of these drawings were later scrutinised, coded and summarised. Literally hundreds of such drawings were produced, but only the drawings of those children who were interviewed individually have been included in these analyses and it should be noted that the sample was slightly depleted by the fact that drawings were not available from all children. (Classroom-based techniques of data collection are inherently more vulnerable to local exigencies.) In the post-intervention interviews, a similar task of describing the origins of some manufactured materials was approached orally, using a piece of cotton thread and a steel rod. Because of these differences in the elicitation mode and the differences in the materials themselves, the pre-intervention responses are presented separately and in full in this chapter. More general comparisons between the two techniques, as well as some comments on the specific responses, are presented in Chapter Five (Section 5.4.1).

Cotton fabric

Table 3.15 summarises children's ideas about the origins of the cotton material. The large number of responses categorised under 'other' include ideas about the material having been cut from a larger piece of material, sometimes a garment, or having been stitched together from smaller pieces. (The post-intervention categorisation was used to classify responses, for purposes of comparison, though the question was slightly different and consequently elicited a slightly different response pattern.) In a similar vein, a small number of children identified where they could locate the manufactured product, i.e. in a factory or shop, rather than its structural origins. It is perhaps important to recognise that this kind of locational information about where materials may be stored or purchased was sufficient for some children. Twenty-three per cent of the sample (including over a third of the lower juniors) traced the origins back only as far as another product having a very similar form, namely, rope, string or thread.

Table 3.15   Origins and transformations of materials: coloured cotton fabric

	Pre-intervention		
	Infants **n=17**	**Lower Juniors** **n=19**	**Upper Juniors** **n=21**
From sheep	6 (1)	32 (6)	33 (7)
Cotton 'tree'	–	–	14 (3)
Cotton plant/grows	6 (1)	–	14 (3)
From rope or string (or thread)	12 (2)	37 (7)	19 (4)
From factory	12 (2)	5 (1)	–
Shop/supermarket	6 (1)	–	–
Other origins	47 (8)	26 (5)	14 (3)
Don't know	6 (2)	–	5 (1)

About one-third of the lower and upper juniors identified the fabric as originating on a sheep, obviously confusing the origins of cotton and wool.

Figure 3.6   Sheep as the origin of cotton

Twenty-eight per cent of the upper juniors correctly identified the cotton as originating from plant material, though half of these described the source as a 'tree'. The non-indigenous cotton plant was not at all familiar to the great majority of children, apparently.

Figure 3.7   Cotton on trees (drawing with teacher annotation)

what it is like	what it was like before that	and before that	and before that
The material feels nice. It is strong. It is square. You can bend it. It can have patterns on it. Sometimes you wear it to keep you warm.	You put the cotton (a long string with fluffy bits hanging out, soft) on a spinning machine. It turns it into cotton without the bits hanging out.	A long piece of cotton. It has got fluffy bits hanging from it. It is soft and it is like wool. If you walk into it it feels like cobwebs.	It is on a tree. It is like cotton wool but thin bit. It feels like silk. It is white. You can put your ha right through it.

As well as the origins, it was possible to discern various manufacturing or other transformation processes in children's drawings. The number of transformations was treated as the number of states minus one. Each frame in a child's picture strip was not treated as equivalent to a manufacturing process or transformation. For example, children might sometimes present a magnified view, or the same material from a different aspect. The drawings were carefully scrutinised to determine what processes had been indicated. Explicit suggestions about processes leading to the production of coloured cotton material were relatively infrequent. Very often, children drew cotton in different states in adjacent frames of their picture strips, yet it was not possible to infer from those states what specific intervening process might have been assumed.

Metal spoon

Only two infants and no lower juniors made mention of ores, rocks, mines or unformed metal under the ground in describing the origins of a metal spoon.

Figure 3.8   Mining for metal

The upper juniors showed much more familiarity with the sources of metal within the Earth's crust; three upper juniors described the original state of metal as being liquid (see Table 3.16).

Table 3.16   Origins and transformations of materials: metal spoon

| | Pre-intervention | | |
	Infants n=17	Lower Juniors n=19	Upper Juniors n=21
From ore or rocks	6 (1)	–	–
From mine/underground	–	–	10 (2)
Unformed metal in or on ground	6 (1)	–	19 (4)
Originally metal in liquid form	–	–	14 (3)
From metal in another shape (or with another property)	59 (10)	74 (14)	38 (8)
Recycled from scrap/from the tip	6 (1)	5 (1)	19 (4)
From a factory	6 (1)	11 (2)	–
Always existed in present form	–	–	–
From another material	6 (1)	–	–
Don't know/no response	12 (2)	11 (2)	–

By far the most common idea was that the metal spoon had been fabricated from another form of metal, either from metal in another shape (67 per cent of the sample) or recycled from scrap materials (11 per cent of the sample).

As when describing the origins of the cotton fabric, a minority of younger children were content to locate the origins of the metal spoon in its manufactured condition within a factory.

Figure 3.9   Recycling of scrap

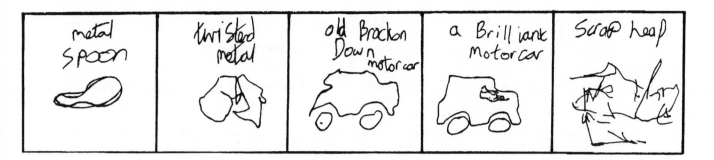

Heating was the transformation process to which children made most frequent reference in their descriptions of the production of metal spoons (11 per cent of the sample), with the juniors more specifically referring to a molten state. About a quarter of the sample explicitly suggested a shaping, rolling or crushing process other than moulding. The proportions quoted should be treated as very conservative estimates since, in the majority of cases, children did not (and were not under any obligation to) specify in their drawings what process had led to the change of form.

3.4.2   Ideas about the possibilities of transforming metal

In contrast to the open-ended question posed about the origins and transformations of materials described in the previous section, children were asked a series of much more focused questions relating to the possibility of transforming metal from one form to another. Specifically, the question was posed whether a metal rod could be transformed into each of four other products. The discussion took place with five specimens available for children to observe: a 10 cm length of steel rod (which was treated as the starting material in each case); a length of wire, metal foil, metal powder (fine filings) and metal wool. Pre- and post-intervention data are available for all except the metal wool, the latter being discussed prior to intervention only.

Responses were analysed by reference to a common set of criteria. Whether or not the transformation was deemed to be possible was the first consideration. If it was held to be possible, a record was made of whether heating or mechanical processes (or both) were assumed. If the transformation was deemed to be impossible, the incidence of three particular reasons was recorded: assumptions that the kinds of *metal* were different (an assumption which in some cases was correct); assumptions that the two end products were quite different *materials*; an explanation describing different *attributes* of the two starting materials as irreconcilable was the third kind of reasoning recorded. Table 3.17 summarises the responses to the question of whether or not it was possible to transform the metal rod into metal wool.

**Table 3.17   Ideas about the possibility of transforming metal:
metal rod to metal wool**

	Pre-intervention		
	Infants **n=23**	**Lower Juniors** **n=23**	**Upper Juniors** **n=22**
Transformation deemed possible, described in terms of:			
Use of heat	4 (1)	4 (1)	–
Use of machine/ mechanical process	4 (1)	4 (1)	14 (3)
Use of heat *and* mechanical process	–	–	14 (3)
No specified method	4 (1)	–	9 (2)
Total suggesting rod to metal wool transformation was possible	13 (3)	9 (2)	36 (8)
Transformation deemed ***not* possible because:**			
The two objects had different attributes	48 (11)	48 (11)	41 (9)
The rod and wool were different metals	–	4 (1)	–
The rod and wool were different materials	4 (1)	26 (6)	14 (3)
Total suggesting transformation *not* possible	52 (12)	78 (18)	55 (12)
No response/don't know	35 (8)	13 (3)	9 (2)

Around one-tenth of the two lower age groups considered that it might be possible to transform the metal rod into metal wool; 36 per cent of upper juniors thought it would be possible. Between 40 and 50 per cent of all three age groups described the transformation as impossible because of the perceptible differences in the forms of the two products. Fifteen per cent of the sample asserted that the metal rod and metal wool were actually different materials. (This was not an assumption that they were different metals; the wool was not recognised as metal.)

The general indications were that children were tending to focus on the superficial attributes of the two materials and lacked the knowledge and experience to make a more abstract underlying link. Both the word 'wool' and the texture and appearance of this item were felt to be a possible distraction which was emphasising a material difference between it and the steel rod. Consequently, metal wool was omitted from the post-intervention replication of questioning in this area.

3.4.3   Predicted changes on heating

This was an area which was explored pre-intervention only; no detailed intervention work on change of state had been programmed, so the ideas and more specific hypotheses which children aired during the interviews could not be specifically investigated. The interviews revealed children's assumptions about expected changes when the metal rod, cotton wool and vinegar were subjected to heating. Since the degree of heating has an important bearing on the resultant changes, this was specified at the beginning of this phase of questioning, as follows:

> *Imagine that we put each of these, the steel rod, cotton wool, vinegar, into a small open container that you can heat to a very high temperature - as high as you like.*

Metal Rod

Responses are summarised in Table 3.18. The largest category of response at all ages was the suggestion that the metal rod would melt or go soft (the latter being a more equivocal reference to a possibly implied change of state), suggesting that children were familiar with the notion of change of state of a metal. How many had actually witnessed such a phenomenon either at first hand or through secondary sources is not known. Some children mentioned experiences of having seen soldering or sequences showing molten metal on television.

54

Table 3.18   Predicted changes on heating: metal rod

	Pre-intervention		
	Infants n=23	**Lower Juniors** n=23	**Upper Juniors** n=22
Will melt or go soft	48 (11)	83 (19)	86 (19)
Will burn/burn away/ catch fire	9 (2)	–	–
Will change colour only	4 (1)	4 (1)	5 (1)
Will get hotter with no other change	4 (1)	9 (2)	9 (2)
Will get harder	13 (3)	–	–
Other	9 (2)	4 (1)	–
Don't know	13 (3)	–	–

An associated factor which was pursued was children's assumptions about the resultant mass of the metal following the heating process. Responses are summarised in Table 3.19.

Table 3.19   **Assumed effects of heating on the mass of the metal rod**

| | Pre-intervention | | |
	Infants n=23	Lower Juniors n=21	Upper Juniors n=20
Weigh more	–	5 (1)	10 (2)
Weigh same	4 (1)	5 (1)	45 (9)
Weigh less	35 (8)	71 (15)	40 (8)
Weigh nothing	9 (2)	5 (1)	–
Don't know	4 (1)	4 (1)	5 (1)
No response	48 (11)	9 (2)	–

Forty-eight per cent of the sample overall predicted that there would be a net mass decrease in the metal rod as the result of heating; five per cent predicted that nothing of the metal would remain. There may be evidence of a trend in which those younger children predicting an outcome of change in mass expected mass decrease while the older children expecting a change in the mass predict an increase. Almost half the upper juniors predicted a constant mass.

> *It would be like liquid. It would be lighter after heating, cos water's light.*   Y2 B H

> *It would melt. It would be all squashed. When its melted it wouldn't weigh anything at all. It's too soft.*   Y3 B M

> *No, it wouldn't weigh the same, it would weigh less. Most of the heaviness would have melted.*   Y5 B M

Cotton wool

There was a strong age trend in the increasing understanding of cotton wool as a combustible material - 39 per cent of infants, 70 per cent of lower juniors and 91 per cent of upper juniors - with about half of each group indicating that there would be a resultant reduction in the material. A quarter of the infants and nine per cent of the lower juniors suggested that the cotton wool would become hotter with no other change.

Table 3.20   Predicted changes on heating: cotton wool

	Pre-intervention		
	Infants n=23	**Lower Juniors** n=23	**Upper Juniors** n=22
Burn and become less (reduce)	13 (3)	35 (8)	50 (11)
Burn, catch fire (no further elaboration)	26 (6)	35 (8)	41 (9)
Get hotter, but no other change	26 (6)	9 (2)	–
Other	30 (7)	22 (5)	5 (1)
Don't know	4 (1)	–	4 (1)

It will get hot. It won't change.   Y2 B M

Nothing would happen to it.   Y2 B M

It would go hot. Nothing else would happen to it.   Y2 B L

Vinegar

A quarter of the upper juniors referred to evaporation of the vinegar on heating, with a further five per cent simply suggesting that it would 'go away' or 'disappear' (see Table 3.21). No children in the younger age groups used the term 'evaporation'.

Table 3.21   Predicted changes on heating: vinegar

	Pre-intervention		
	Infants **n=23**	**Lower Juniors** **n=23**	**Upper Juniors** **n=22**
Become hot, no other change ('stay the same')	35 (8)	57 (13)	27 (6)
Go away/disappear	13 (3)	4 (1)	5 (1)
Go into air/evaporate	– 	– 	27 (6)
Other	22 (5)	30 (7)	23 (5)
Don't know, no response	30 (7)	9 (2)	18 (4)

The incidence of any reference to boiling, bubbling or foaming as the result of heating was also recorded. This might be assumed to be a more 'everyday' experience than evaporation; reference to this phenomenon increased steadily with age to reach 36 per cent amongst the upper juniors, with nine per cent of infants and 22 per cent of lower juniors mentioning this phenomenon.

A surprisingly large number - 40 per cent of the sample - indicated that they assumed the vinegar would 'stay the same' apart from the fact of getting hotter.

58

4. INTERVENTION

4.0 Planning for Intervention

Before meeting with Project teachers, the research team was able to make a preliminary survey of children's ideas. This review, prior to the full data analysis, was based on experience of interviewing children, evidence in written form collected during the exploration phase and conversations with teachers during this phase. This survey was necessarily brief but it nevertheless sketched a framework on which intervention could be built. Intervention is a phase of the research in which teachers respond to children's ideas with the intention of encouraging them to take those ideas further. To do this, teachers have to be aware not only of children's starting points as ascertained during exploration but also of lines of development to undertake. Four areas in which children's ideas could be developed correspond to the four issues described in Chapter One.

Firstly, children's descriptions of materials, though rich and varied, suggested a need to develop ideas about properties. Some children, for example, had focused on use or other associations when describing materials while others had indicated what the materials were like or how they behaved. While some characteristics appeared to be clearly defined in children's thoughts, others such as concepts of strength and hardness had more blurred boundaries. Children's attempts to find out more about materials had sometimes lacked clear purpose. One concern was thus to encourage a more systematic approach.

The second area of development related to children's concepts of solid, liquid and gas. Although physical state is only one particular property of materials, it was felt that it was sufficiently discrete to pursue as a separate area. Some children had found difficulty in generating examples of liquids; even greater difficulty was experienced with the term 'gas'. Similarly, while a particular solid, steel rod, and a particular liquid, vinegar, had generally been accurately categorised, other examples had produced a greater variety of response. It was therefore of interest to see whether further experience might help children to develop more accurate and more general ideas about what is a solid, what is a liquid and what is a gas.

The third area for development was in terms of ideas about the uses of materials. Some children had been able to give reasons why particular materials had been chosen for particular purposes. Reasons were expressed in terms of functional properties of the material or in terms of manufacturing properties or, more rarely, were based on either aesthetic considerations or, more rarely still, on economic factors. It was thus desirable to see whether further experience would lead to the development of more complex multifaceted ideas.

Finding out about some properties, such as colour, texture or mass, does not necessitate changing a material at all. The examination of other properties does, however, change the material in some way. Examples include changing shape when comparing softness or flexibility, adding water to test solubility and changing state through heating and cooling. Accepting that materials can be changed would appear to be a prerequisite for being aware of origins of materials. That is, a child who does not believe that a material can be changed into another form or, at times, into another material cannot be aware of where the new form or new material comes from. Thus, a fourth area for development involved considering where materials come from and the changes that occur during their manufacture.

It was thought desirable that, during intervention, teachers should attempt to promote development of ideas in each of these four areas. The time available for intervention was approximately five weeks. This meant that it would not be possible to develop each area extensively. In particular, there was a wide range of properties available for investigation - flexibility, hardness, solubility, strength, transparency, 'weight', volume, compressibility and effects of heating. It would not be possible to cover all of these with each group of children. Moreover, in practice, teachers found that children became engaged in particular areas of work, leaving little time for intervention in other areas. This reflected an overriding principle of the Project philosophy, that teachers encourage children to develop their own ideas.

Teachers and researchers working on previous phases of the SPACE Project had developed a number of strategies for intervention. These were considered in the design of activities for the intervention phase. Strategies included the testing out of ideas by practical investigation, generalisation to help in recognition of instances of a concept and vocabulary work which involved encouraging children to develop more precise meanings for words through concrete examples and actions.

Furthermore, the important effect of interaction with other people's ideas on each person's own concepts was recognised. This interaction could take place between a person and secondary sources such as books. Discussion and debate would provide a more direct and mutually interactive situation.

At the pre-intervention meeting teachers sought to use the strategies described above in designing activities for intervention. They also incorporated their own experience of the kind of activity that would be suitable for children in their classes. Thus, after the preliminary findings of the exploration phase had been outlined, groups of teachers and researchers set about making suggestions for intervention activities. These suggestions were then reduced by further discussion to a number felt to be manageable within the five-week period allocated for intervention. Those activities served as intervention guidelines. They did not represent a common intervention which all teachers were expected to follow. Rather they gave an indication of the kind of activities that might arise as a consequence of encouraging children to develop their ideas. Teachers were aware of the need to respond to the ideas that had been expressed in their own classes. Clearly, the actual activities undertaken could depend upon children's own suggestions. Likewise, teachers would use their experience in trying to match experiences to the individuals or group concerned. The intervention guidelines, shown in full detail in Appendix VI, were a summary of the ideas about intervention suggested at the meeting. Their function was thus to provide a framework to help guide teachers about the direction intervention might take. The actual course of intervention would vary according to the needs and nature of the class. In order to keep a track on what happened in each class, teachers were asked to keep an intervention diary. An outline sheet for this is shown at the end of Appendix VI. Teachers made brief notes on how the intervention activities were tackled. They also recorded comments and reactions of children. Where writing or drawings were produced these also gave an indication of the work in which children were involved as well as occasional insights into how their ideas were changing.

During the exploration phase, teachers had accumulated many materials with which children had worked and could continue to do so. To add to their own resources, each school was provided with a purchased 'materials kit'. Items of this kit could be used where necessary. It included tiles, rubber, wool, cotton and various kinds of metal and plastic. Teachers were also given a list of firms from which secondary resource material could be obtained.

In the following sections of this chapter, each area targeted for development is taken in turn and intervention undertaken in the area is reported. The intention is to give some indication of the kinds of activity that took place together with some of the outcomes of those activities. It does not reflect the work of any particular class but rather, through describing some of what happened in different classes, presents an overall picture of intervention. What was possible for each individual teacher to tackle was necessarily only a fraction of the spectrum of activities described over the following pages.

4.1 Intervening to Develop Ideas on the Properties of Materials and Ways of Describing Them

Work within this area fell into two categories. Firstly, under the general heading of 'vocabulary work' children were encouraged to describe and compare materials. Secondly, under the general heading of 'finding out more about materials' children were helped to devise ways of investigating specific properties of materials. Appendix VI contains the activities. Activity 2a and 2b pertain to the first category while Activity 4a relates to the second.

4.1.1 Describing materials

Two specific suggestions were made for vocabulary work but teachers were encouraged to use those only as examples and also to adopt other ways of enriching children's descriptions of materials. The two suggestions were entitled:

a.   Using sense of touch;
b.   Guessing game.

The first activity involved children being blindfolded and being asked to feel materials. They talked about what they felt. This activity contrasted with the more open activity of the exploration phase in which children were free to describe materials in any way they wished. The sense of touch had not been extensively used during that phase and it was hoped that by focusing on one sense, there would be opportunities to develop children's language. In a similar way, it would have been possible to limit observations to what could be seen or what could be heard.

Teachers generally felt that the 'blindfold' game was more appropriate for younger children. Here are some comments made by different members of a group of infants about the materials which they were feeling but could not see:

> *Pastery, I mean watery. Gungy thing. It's thick.* (tomato sauce)

> *It's flowing like sugar. It's smooth. It's one thing. I can move it round the pot.* (salt)

> *It's wet. It's cold. It's like a river. It's a liquid. It's like it goes all round my fingers, it moves.* (soft drink)

> *It's a bit lumpy. It's one big thing. It's rough. It's a hard thing. It's a stone.* (stone)

Each child has been able to mention more than one characteristic. Some of them have used similes to indicate what the feeling reminded them of. 'Hardness', 'smoothness', 'wetness', 'thickness', 'hotness' were among the properties mentioned and the invariant composition of some materials has also been commented on.

The teacher of these children had earlier noted some confusion in the use of the terms 'soft' and 'smooth'. This had happened when children had described materials after touching them. It is perhaps through direct comparison of feeling a surface and pressing into it that the distinction between texture and hardness can become clearer to children. Another teacher also noted this interchange in the usage of 'soft' and 'smooth'. One child, for example, described wood as 'hard and a bit soft'. 'A bit soft' was a reference to the smooth feel of the wood. After the children of this class had described materials, they were asked to sort them into soft and smooth sets. Classification was thus used as a means of focusing on the meanings of words. In a similar way, another teacher got children to look at different instances of smooth materials, both hard and soft ones. She still reported, however:

> *The children know the two opposites rough/smooth, but still think of soft as smooth in some cases.*

The second kind of vocabulary work suggested was called 'guessing game'. One child chose a material and gave one piece of descriptive information at a time to others in a group. At each stage, the others could use the information accumulated to try to work out what the material was. The child who was describing was asked to think about what the material was actually like rather than mentioning its use, what it was contained in or other information associated with it. It was imagined that this activity would give opportunities to guide children towards accurate use of terms which describe properties. It might also help children to think about which characteristics most clearly define a particular material and distinguish it from others.

Both younger and older children attempted the guessing game. Younger children often found it difficult to restrict themselves to properties and went on to give uses. For instance, one child indicated that the material (metal foil) was 'silver in colour' and then went on to say, 'You wrap things up in it'. The following extract shows the game converted into question and answer form by one group of upper junior children:

Is it liquid?	*No.*
Can you drink it?	*No.*
Is it hard?	*No.*
Can you eat it?	*Yes.*
Is it honey?	*No.*
Is it flimsy?	*No.*
Is it tomato sauce?	*No.*
Is it coffee?	*No.*
Is it powder?	*No.*
Is it washing up powder?	*No.*
Is it salt?	*No.*
Is it powder paint?	*No.*
Is it flour?	*Yes.*

The analysis of such a dialogue with those children involved may help them develop their ideas. Instances where children had not paid heed to previous information could be discussed. Moreover, the characteristics that they have used could be pointed out. Strategies for arriving at the correct material as quickly as possible could also be discussed. This involves thinking about important distinguishing properties and starting with more general properties first.

One teacher helped upper junior children to do this by asking them to group materials according to their characteristics. One child in that class grouped a number of materials into the following sets: 'soft', 'squashy', 'flexible', 'smooth', 'transparent', 'shiny' and 'bitty'. Some materials were included in more than one set, showing a recognition that a material can possess more than one characteristic. Some of the set labels and some of the classifications might need further discussion; for example, 'What is the difference between *soft* and *squashy*?' Nevertheless, some of the characteristics chosen could be turned into questions for the guessing game. 'Is it bitty?', for example, would be a useful question to ask after it had been ascertained that a material was a solid.

Further examples of attempts by teachers to encourage children to think in terms of different properties are shown below (Fig. 4.1 and 4.2).

Figure 4.1

Looking at Materials.	A (lid)	B (wool)	C (cotton)
feel like	Rubber	soft	soft
colour	Black	Cream	bage
shape	Cirle	Cloud	Clound
patterns	Spokes	string	Curled lines
smell	Old house	Cow muck	Doesn't smell
heavy / light	light	light	light
size	4×4cm	Any	Any
What happens if.... bend it	It flips back	Nothing	Stretchs
Scratch it bounce it	Makes a noise	it ripes Doesn't	makes a small noise Doesn't
twist it	Bends	Twists	Twists
pull it	Doesn't	Comes a part	Comes a part
rub it.	Bumpy	stringy	soft

Figure 4.2

MATERIALS	SAME	DIFFERENT
Say what you	have done to the materials to find	out this information
brass and Zinc	They are not the same I looked	The Zinc is sharper I felt and They are diffrent and looked colours
brass and aluminium	They both bend I bent them	They are diffrent colours I looked

The first example (Fig. 4.1) shows a structured framework introduced by the teacher after discussion with the class about the kinds of things they might notice about materials. In the second example (Fig. 4.2), children are less restricted as to the kind of observation they can make although they have to search for similarities and differences. In both instances, it is to be noted that some aspects of description involve interacting with the material in some way - seeing if it bends and so on - rather than simply observing. In this way, description and comparison of materials cannot rely on observation alone but necessitate the practical examination of particular properties.

4.1.2   Investigation of properties of materials

The guidelines for the practical examination of properties of materials by children are given in Activity 4a (Appendix VI). It was suggested that children find out more about materials by raising questions which they could investigate. As illustration, the following examples of questions were quoted:

> *Which piece of metal is the bendiest?*
> *Which materials can I see through?*
> *Is this piece of wood harder than this one?*

Teachers were asked to encourage questions from children wherever possible. Properties which could be investigated included transparency, weight, volume, strength, hardness, solubility, flexibility and compressibility. This kind of activity was similar to Activity 2 of the exploration phase in which children interacted with materials using implements of their choice. The difference lay in the more open nature of the exploration phase activity in contrast to the more structured form of this phase. Thus, while groups of children were expected to plan and carry out investigations themselves, teachers guided children by, for example, advising on the selection of materials, by encouraging them to think about what they were testing and by prompting them to consider to what extent a test was a fair one.

There appear to be three particular difficulties to be faced in these tasks. The first is in deciding what a term such as 'transparent' or 'hard' meant. The second is to operationalise that meaning into something that would test it. Having decided, for example, what 'transparent' means, how can transparency be tested? The third is in making a fair comparison. In practice, the first two issues were not ones that children directly addressed. They tended to seize on some manipulation of the materials without considering whether that manipulation was pertinent to the property they had in mind. Indeed, the urge to interact with materials almost seems to preclude consideration of 'What does a given *property* mean?', and 'What do I have to do to make comparisons of the *property* for different materials?' Some properties such as transparency and 'weight' appeared to be more readily understood and led to accurate means of testing. In contrast, properties such as flexibility, hardness and strength did not so readily evoke a means of testing. Where they did, they were often confused. Three children who had asked 'Which material is the hardest?' decided on brass because it would not break. Another three who had asked 'Which material is the softest?' chose raw wool because it was easy to break. Hardness had been equated with 'breakability' and thus had not been differentiated from strength.

Sometimes, the materials themselves seemed to inhibit children from devising investigations. This was the case for a group of lower juniors who wished to find out which material was the 'bendiest'. One of their materials was a piece of rubber which bent under its own weight. There was clearly no need for a test; the rubber looked the 'bendiest'. Moreover, it is likely that children's previous experience had given them fixed ideas about what kind of materials were bendy. It was only when they had to compare two pieces of metal which did not obviously differ in 'bendiness' that the group of children really had to do something with the materials.

They then made two suggestions. One was to press the material at both ends to see if it would bend. The other was to hang the material over a desk and press it. In both cases, they were judging the ease with which something would bend. That is, they were feeling the force they needed to use to bend it. They were deciding which bends more easily (with the least force) rather than which bends more. The latter offers a greater potential for accurate comparison because the amount of bending can be measured. Instead of estimating by feeling the force needed to bend, the force can be kept constant and the amount of bending compared. (These children were not familiar with the use of forcemeter). This is what one group of infants did (Fig. 4.3).

Figure 4.3

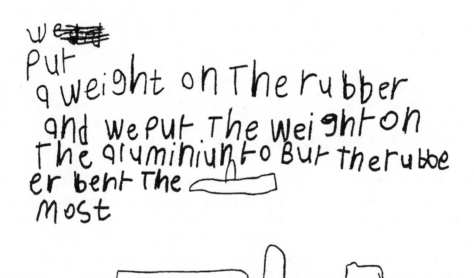

The test had been made fair by putting the same weight on each material. However, it was not clear where the weights had been placed on each material and how the amount of bending had been judged. A further difficulty in comparing materials was pointed out by the group of lower juniors who had compared which piece of metal bent more easily. To make their test fair, they had wanted pieces of material of the same shape and dimensions. The difficulty of obtaining equivalent samples means that some fair tests are only of particular items rather than of the material of which they are made.

That is, the test may show that a particular piece of rubber bends more than a particular piece of aluminium rather than rubber bends more than aluminium.

The following comment by a teacher indicates the general reaction of children to this work:

> *They found the whole exercise difficult, wanting to test more than one thing at once, not being sure about fair testing and writing up was not very clear. For testing 'hardness' they tested 'brittleness'. Nevertheless, as there was much discussion I feel these were useful sessions and later testing should prove more successful.*

Teachers responded to children's ideas in ways appropriate to the kind of thinking expressed. Some young children, for example, were keen to explore materials by manipulating each one. They did not appear to be ready to compare materials one by one and so were asked to divide them into two groups; those that bent and those that did not bend. Here is the record that one child made of this (Fig. 4.4).

Figure 4.4

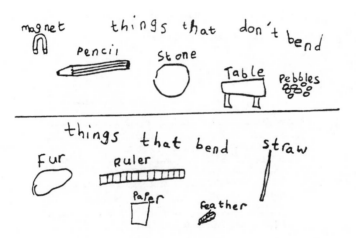

Although no comparison of the degree of bending was involved, children had made a start in comparing materials. Through this experience, some began to recognise different amounts of bending. 'It went to my thumb' explained one, indicating how far a piece of rubber had bent. Others began to express, in their own terms, the need to control variables 'He's stronger. He's pulling too much', was a comment made by a member of a group bending some materials.

In another investigation of bending with upper juniors, the first suggestion was to use a stopwatch to time how long it would take to snap each piece of metal. When they were asked to ensure fair testing, they indicated that the same person should do the bending each time. They were then asked to consider whether that person would be pressing each material in the same way. This led to the suggestion of stamping on the material. On realising that a 'stamp' would not necessarily be constant, some suggested adding weights to the material. This led to a redefining of the investigation and a realisation that the first suggestion would not determine which bent the most but which snapped under the least number of 'bendings'.

68

This example illustrates some of the difficulties children had in matching what they were going to do to what they thought they were finding out. It also illustrates the development of their thinking about 'fair testing'. It was quite common for children to suggest, in the first instance, that being fair meant that the same person should be involved.

The examples of investigations into bending have been discussed at some length to give some indication of ways in which different children's investigations developed. This was not the only property which children suggested for testing. Different groups of children carried out investigations into strength, hardness, 'weight', solubility, absorbency, staining, effect of water on metals, transparency, conduction of electricity and whether materials floated or sank. The following three examples, all written by upper junior children, give an indication of the scope and extent of this work. It should be remembered, however, that each individual child is likely to have had an opportunity as part of a group to investigate only one property. Moreover, these records are initial plans or final products of work and, as such, cannot reflect the processes by which the investigation was adapted until it took on its final form. Some investigations may have been abandoned along the way and less successful attempts may not have been reported.

The three examples are records of investigations of 'weight' of different liquids (Fig. 4.5), strength of fibres (Fig. 4.6) and absorbency (Fig. 4.7).

Figure 4.5

liquids	weighs	Result
water	36 grams	
Tomato Sauce	42 grams	
cooking oil	33 grams	
orange Juice	36 grams	
coke	38 grams	
Salt water	45 grams	
white paint	47 grams	
milk	50 grams	
Fabric conditioner	40 grams	

weighing liquids

At first we collected some liquids to measure. And we measured 30ml of water and other liquids in a small container. We used the same container each time, and we weighed the 30ml container of each liquids to see the results. We have found out that all liquids weigh diffrently

The cooking oil was the lightest because it floats on water, the milk is the heaviest we think because of the calcium.

The children whose investigation is shown in Figure 4.5 did not only weigh liquids, but also saw a need for controlling the volume of liquid used. Although individual results may not be that accurate, the children have been able to conclude that liquids do vary in respect of this property, the 'weight' of a set volume. The speculations as to the reasons for these differences are also interesting. The low value for the 'weight' of oil has been associated with its floating on water.

Figure 4.6

> Experiment.
> To find out the strongest fibre.
> What we did
>
> We pulled out one fibre from each piece of material. We stuck it to the table with sellotape. Then we tied a loop on the bottom of the fibre and then we put an S-hook through the loop. Then we got some string, tied it to a tub. Then we put a small nuber of weights in the tub and kept adding until the fibre broke we used the same tub each time to make the experiment fair we wrote down the breaking weight.
>
> | Name of Fabric | Weight Breaking string in g |
> | --- | --- |
> | Linen | 1035g |
> | wool | 7g |
> | silk | 180g |
> | nylon | 48g |
> | cotton | 16 g |
> | Acrylic | 241g |
>
> What we found out
>
> We found out that linen was the most strongest fabric and wool was the weackest fabric. It might of been diffrent if they was the same thickness and size, they mite of been diffrent.

Figure 4.6 is a report of an investigation into strength of fibres. A valid criterion for judging strength has been chosen - the breaking of the fibre. Moreover, by adding weights, it has been possible to compare by measuring rather than simply by feeling a difference in how easily the fibre breaks. The final statement in this report shows an appreciation that it had not been a fair test of the materials themselves but only of those particular fibres.

Figure 4.7 is a plan for an investigation of absorbency. The children who devised this plan seem to have a clear concept of absorbency and have thought of a way of assessing it by measuring weight before and after. They have thought about how to make it a fair test; they were going to immerse the material for the same time and to shake the same number of times to remove non-absorbed water. The children made predictions about which materials would absorb water and which would not. The testing of these predictions allowed the children to develop their ideas about the materials whose porosity they had not judged well.

Figure 4.7

Investigation: Absorbency of materials.
What we are going to do.

We are going to see which materials absorb water and which
do not absorb water. We are going to use a tissue, a lump of
plasticine, a small peice of lined scrap paper, a metal spoon, a large
peice of corrugated cardboard, a plastic stickle-brick, a brick,
bolsa wood, rubber, paint and raw wool. We are are going to see
if the material absorbs any water by weighing it and putting
it into water for a minute. When we take it out of the water
we will shake it four times to get rid of any water on the
surface and then weigh it again.
What we think is going to happen.
I think the, bolsa wood, tissue, plasticine, rubber, paint, paper
and raw wool. I think the plastic will not absorb water
along with the metal spoon and the brick.

4.2   Developing Ideas about Solids, Liquids and Gases

This particular property of materials, their usual state as solid, liquid or gas, had been chosen for more intensive study. The intervention guidelines can be found in Activity 2c in Appendix VI.

The guidelines gave only a very general outline of the kind of activities that might result in development in this area. This outline reflected three main principles: exemplification, generalisation and awareness of characteristics. Children were encouraged to find examples of liquids, say. These might be instances they had observed around them. They might be from pictures they had seen. By bringing in real examples or representations of them, a display could be formed. This display would form a focus for thought and discussion. By asking questions about the items, children might be encouraged to form more general ideas about liquids. It had been noted, for example, that colourless liquids had been treated as water by some children. They could be asked to find, or think of, other colourless liquids. In other words, they were being encouraged to generalise so that they might discover whether their idea about liquids seemed restricted in some way. Thus, finding examples -

exemplification - was followed by extending those examples, that is, generalisation. At this stage, children might be ready to make explicit how they were recognising liquids, for example, by talking about what they were like. Their ideas could then be tried out by asking them to investigate the characteristics they had mentioned. This would be done with the intention of raising their awareness of the characteristics of particular states. Rather than developing any formal definition of a particular state, children would be guided to develop good describing words for it.

4.2.1   Solids and liquids

Using their experience of the exploration phase, teachers of infant groups judged that intervention on gases would not be productive. They concentrated on solids, and liquids, direct experience of which could be more readily given. Here is an account by one teacher of children's experience with different materials:

> *I asked various children to put first Lego, then buttons, then washing up*
> *powder and fourthly salt into a large, empty jar. We all listened to the*
> *noise each one made and agreed that even Lego could be poured from one*
> *container to another. Each jar was tipped to one side and children*
> *described what had happened. 'It's gone into a heap', 'It's made a wall in*
> *the corner'. Then each jar was carefully restored to an upright position,*
> *many of the children predicting that the substance would spread out again.*
> *The next things to be poured out were water and orange cordial. Children*
> *predicted that this mixture would not stay in the corner when the jar was*
> *righted.*

These children were able to distinguish between materials which remained in a heap and those which spread out. One child explained and drew it as follows (Fig. 4.8).

Figure 4.8

if you pour a liquid into a Jar it will go straight to the bottom. if you pour a solid into a Jar it will go vertical

Liquid honey

Solid BRICKS

This child had also been able to attach the labels 'solid' and 'liquid' to the two groups of materials.

Another teacher of infants used the materials which had been on display in the classroom during the exploration phase and asked children to sort them into solids and liquids. She reported:

> *Children knew some things were solids straight away, for example, wood, sponge, cotton wool, but were unsure of foil, polythene bag, salt and coffee. For liquids children readily used words like 'runny', 'thin', 'thick', 'You can pour it easily but sometimes it's too thick to pour', 'It takes a long time to move like honey', 'You can roll it around'. I've set up a liquids table.*

It would appear that these children had been able to recognise liquids but had had difficulties in categorising some kinds of solid. Other teachers also provided examples of materials which challenged category boundaries. For one group of children, a discussion started up about flour. Some of them maintained that flour was not a solid, even though it could also be neither a liquid nor a gas. The flour did not behave as other solids did. Children were encouraged to look at the flour more carefully and they decided to examine it under a microscope. They then said that the flour was 'lots of little solids'. That is, although the flour as a whole behaved in a way atypical of solids, each individual bit resembled other solids sufficiently to enable it to be categorised with them.

Another teacher addressed children's uncertainty about the status of powders by getting them to make powder from a more apparently 'solid' form of the material. A stick of chalk was used and the bits left behind on the board were examined. Accepting the stick of chalk and the powder as the same material could lead to categorising them in the same way, that is, as solid. It is still possible, of course, for someone to believe that the transformation in the size of fragments also transforms the state.

Some of this same group eventually arrived at the idea that something was a solid if it was not 'watery'. They talked also of sweets and chocolate going 'watery' when hot. They appeared to treat water as the representative liquid. The analogy with water could become misleading; melting does not produce water. However, it is possibly a useful starting point if children are totally unfamiliar with the word 'liquid'. Children might be asked, for example, to group materials into 'watery' ones and others.

Children often spontaneously gave their reasons for calling an item a solid or a liquid. On other occasions, teachers asked 'What makes you say that?' so that children would justify their assertion. Whichever was the case, children might have been challenged to consider whether the reason was a sufficient one. In other words, was a reason given for X being a solid, characteristic of all solids and, at the same time, did that reason exclude non-solids? As an example, one of a group of children had said, 'Coke is a liquid because you can drink it'. They were asked if they could think of any liquids that could not be drunk and realised that not all liquids were beverages (at least not safely so!). Another said that the coke was a liquid 'because you can pour it'. The group then were asked to examine a number of materials for 'pourability'. They reconsidered the idea when they were able to pour some solids, particularly powdery ones. They then set about comparing the way solids and liquids poured and noticed the different way in which liquids and solids flowed as they were poured. The first child's reason, 'drinkability', had been too narrow in excluding some liquids while the second child's, 'pourability', had proved too broad in including some non-liquids and had needed refining.

Items of mixed composition were also used to provoke discussion. A group of infants eventually decided that lemons were both solid and liquid. Similarly, changes of state were examined by some groups. A mixed age class decided that correcting fluid ('liquid paper') was a liquid in the bottle but dried as a solid. An upper junior group was asked to think of conversions between solid, liquid and gas. Here is one product of that work (Fig. 4.9).

Figure 4.9

Interestingly, of the two liquid to solid changes quoted, the first is an example of solidification by freezing while the second is an instance of 'drying out' by evaporation.

Another example of developing understanding through a focus on change of state comes from an infant class. The children watched the teacher making scrambled eggs. Most of them said that the egg was solid although one child said it was 'solid outside and liquid inside'. As cooking proceeded, various comments were made about the butter, the melted butter, the milk and eggs until the final product was adjudged to be a solid. Children then thought of other examples which were similar to what had happened to the butter and compared the amounts of heat needed for those changes. They said that, compared with the butter, snow needed much less heat to melt while metal needed much more.

Many teachers asked children to produce examples of solids, liquids and gases. The actual examples or pictures of them were displayed and formed the basis for discussion. A lower junior group produced the following set of words to describe the liquids in front of them: 'bubbly, thick, slimy, juicy, cold, greasy, mushy, poisonous, splashy, sticky, gooey, runny, smelly, wet, thin, watery, tasty, colourful, hot'. Through discussion of these qualities children began to develop an understanding of what liquids could be like. An upper junior group took this a step further and classified liquids according to their own criteria. The following table was drawn by one child to show how this had been done (Fig. 4.10).

Figure 4.10

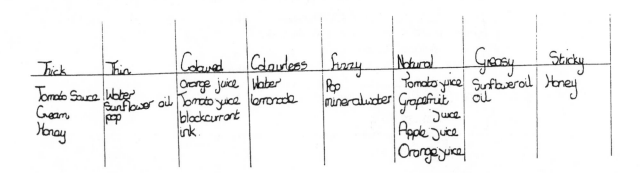

Some liquids appear in more than one column reflecting the fact that the chosen categories are not mutually exclusive. Children then went on jointly to produce a collage with the describing words attached. In a similar exercise with solids, the adjectives chosen by one child were 'sparkling, shiny, hard, soft and rigid'. The 'rigid' list included cooking foil and tree along with brick, tile, concrete pillars and wall. This provided an opportunity for discussion of the child's meaning for the term 'rigid'. Some children noticed the overlap of categories and produced this Venn diagram (Fig. 4.11).

Figure 4.11

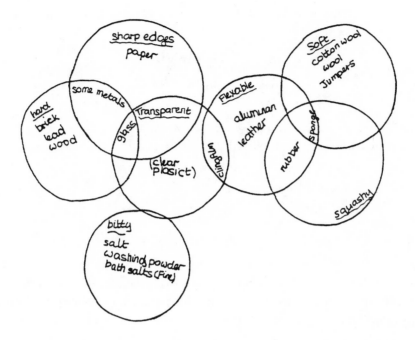

Again, this provided opportunities for discussion of which descriptive terms could overlap and which excluded one another. Questions like 'Could something be hard and flexible?' and 'Could something be bitty and hard?' could lead to a change in ideas about these properties.

A further variation on this theme of looking at different types of liquid and solid was taken up by another upper junior group which started by listing characteristics of different solids. Sometimes these were expressed as opposites such as firm/soft. Others were given as individual qualities such as sharp, metallic, magnetic, dissolves, hollow, granulated, powdery, fibrous, man-made and reflective.

They then went on, with varying degrees of success, to make keys for identifying a set of solids. Figure 4.12 is one example of a framework for such a key.

Figure 4.12

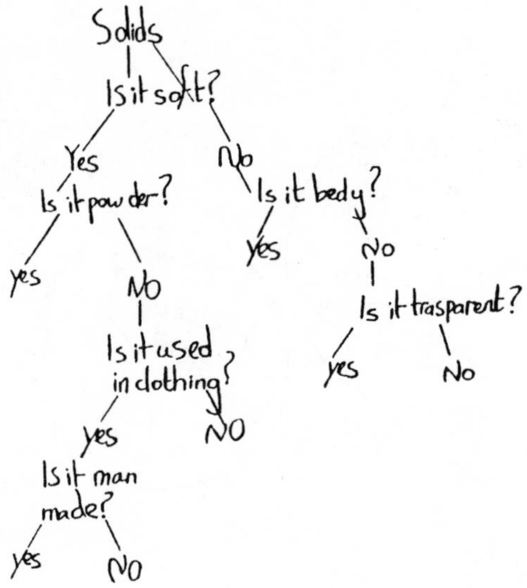

The forming of a key for some solids is not an easy exercise and the example shown is only partially successful. It is difficult to find a gradation from more general properties with which to start to more particular ones.

What the production of descriptive terms, set lists, Venn diagrams and keys all have in common is that they help identify the characteristics of different kinds of solid or liquid. Through recognising, for example, that solids can be hard or soft, children can be brought to a greater awareness of the range of items that count as solids. They might be less likely to characterise solids as hard. While some teachers had adopted the strategy of encouraging children to look for characteristics common to all solids or to all liquids, these teachers had also drawn attention to the existence of different characteristics possessed by different groups of solids or of liquids. Both looking for similarities across the whole set and looking for similarities of groups within the set had potential in helping children draw concept boundaries. In the latter case, children might be developing their notions not only of solid and liquid but also their ideas about the properties of those materials. In effect, development was possible within two of the areas referred to in this chapter - that of the present section (4.2) on the states of matter and that of the previous section (4.1) on properties in general.

4.2.2   Gases

Work on gases was more limited and was generally restricted to junior aged children. One teacher of infants did, however, use an activity which involved a gas although the term itself was not used. Children in that class looked at the changes when egg whites were beaten.

> *The reception children were quick to describe the changes, 'It's gone from yellow to white'. 'There's more now'. Year 2 children said that the bubbles were making the egg whites bigger. We tried blowing the beaten whites and recognised that it was light and could be blown away. No children were able to surmise that air had been introduced and caused the changes.*

It seemed that, although bubbles were readily recognised, they were not linked to air. Whereas water may sometimes appear to be treated as the typical liquid - liquids are 'watery' - air is not so central to children's ideas about gas. Indeed, there was some evidence during exploration that 'gas' can evoke an image in some children's minds that positively excludes air. That is, it is a 'special' material that comes out of pipes and cylinders in cookers and heaters.

Some upper junior children were asked to generate examples of gas through discussion with one another. They then portrayed some of those ideas by drawing. Here is an example (Fig. 4.13).

Figure 4.13

The drawing starts off with home-based examples, gas from a cooker and from a heater. A cutting device fuelled by 'gas' is also drawn. The other examples do show air but in all cases, the air is bounded by something: by a balloon, by tyres and even by the unnamed air which surrounds the air squeezed from under the water skis or from the exhaust. Perhaps, however, it is the difficulty of representing the pervading, surrounding air that precludes its representation.

Another group of upper junior children made a collage of pictures related to gas. This included aerosols and sprays, cookers and heaters, hair driers and fans, the exhaust of a vehicle, a football and an airship. Some of these items provoked further discussion as to whether the examples were truly gases or not.

Other groups had more direct experience of gases. They inflated and deflated bicycle tyres. They looked at lemonade and other fizzy drinks and thought about what the bubbles were. They considered what happened when an aerosol was used or a bottle of perfume was smelled. They blew bubbles and put water on bicarbonate of soda.

One group followed their experiences by thinking of questions to ask about the gas examples they had generated. These included:

> *Can we breathe it?*
> *Could it put your life at risk?*
> *Is it flammable?*
> *Can you see it?*
> *Can you see its effects?*

Through asking this sort of question, they became aware of the variety of gases and the possibility of different gases having different characteristics.

4.3 Developing Ideas about Uses of Materials and their Relation to the Properties of these Materials

During the pre-intervention interview children had been asked for their ideas about why particular materials were used for making particular objects. They had been asked to say, for example, why wood was a good material for chairs.

In the intervention phase the intention was to follow this up through group work. The interviews had taken place with individuals and a range of ideas were revealed. These had been classified as falling within functional, manufacturing, aesthetic or economic spheres. Group work would allow an interchange of ideas which would perhaps lead individuals to reconsider their own thoughts, with the potential for expansion and modification of ideas.

One possible activity involving group work evolved during the pre-intervention meeting. This was entitled 'Materials in use around the school' and appears as Activity 1 in Appendix VI. Children in groups were asked to try to identify different materials in use in and around the school. These could include materials used for construction as well as others. Children might form a table matching a material to its use. This table could then be used as a focus of discussion about why each material had been chosen for its use. Those working in groups were encouraged to come to an agreement about the reasons that made a material suitable for a particular use.

4.3.1 Uses of materials

Two approaches were taken to linking materials to their use. One involved children looking at different objects and saying what they thought the objects were made of. The alternative approach involved children thinking of different kinds of material in the first place and then looking for objects which were made of them.

Thinking about what objects were made of was particularly useful in revealing the kinds of materials that children had difficulty in identifying. Infants in one group had no difficulty in saying what pipes, taps, door handles and coat hooks were made of. 'Metal' seemed to be a type of material they readily recognised by its appearance. Indeed, when faced with an object whose appearance did not immediately reveal its metallic nature, these children were able to put it to a test. Having initially said that a post support was plastic - 'it looks like plastic' - they hit it and decided that it was metal.

In contrast, when that same group came to some ceramic tiles and the classroom sinks, the comments made included:

> *I think it's made of glass. You can tell by the noise.*
> *They're glass but different colours.*
> They're stone [pointing to the sinks].

These children had linked these objects to materials they knew about and to which the objects bore some resemblance, glass and stone. Ceramics would thus not appear to be a separate category of material for these children although some in another class referred to the tiles as being 'pot'. Although the former children had named the material incorrectly, 'ceramics' in a broader sense at an advanced level would include certain kinds of glass and even rocks as well as the more typical bricks, pottery and other varieties of fired clay. The teacher of this group tried to help children develop their ideas about ceramic materials by letting them handle wet clay and then make models with it.

In the second approach children thought of materials first. One group came up with glass, wood, metal and plastic. The following drawing shows how one child in the group recorded the items made from each material (Fig. 4.14).

Figure 4.14

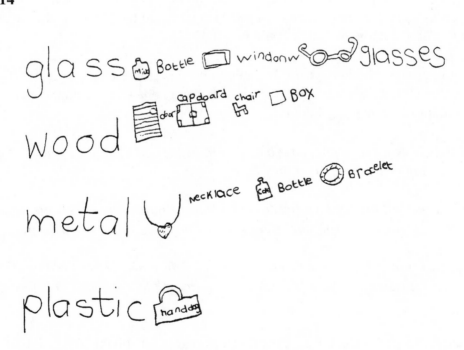

This shows a good recognition of these four materials, while, at the same time, giving a basis for further intervention work. For example, teachers could encourage children to look for other plastics to see if rigid plastics would be recognised. Similarly, leather and plastic handbags might be differentiated.

Many children attempted to identify the material of an object from its appearance. They were looking at and perhaps occasionally feeling objects and listening to the sounds that they made. In another class they took this a step further and used a magnet to help them identify materials. That is, they found out what material was present by testing rather than relying on observation alone. The children were shown samples of different metals including steel, aluminium, brass and silver and told what they were. They were allowed to examine the different samples and found the steel to be the only magnetic metal. Here is what the teacher wrote about the rest of the activity:

> *Only one child was able to volunteer a use for the steel. He said cars were made of it. We went outside and the children discovered that practically all my car, underneath, under the paint and other shiny metal like wheel centres were all magnetic and therefore probably steel. They also found grid covers, dustbins and windows to be magnetic and used their magnets to find out whether the window frames were steel or wood.*

The appearance of different metals is often so similar (and can be masked by protective layers like paint) that the value of using an aid in identification, a magnet, became apparent to these children.

4.3.2   Relation of use of materials to their properties

Identifying materials is, however, only a necessary precursor to the main purpose of intervention in this area. That was to see if children could be encouraged to think about why particular materials were used in particular ways. That is, could they be helped to relate the characteristics of a material to how it could be used?

By working in groups, children were often able to produce more than one reason for using a particular material. These were among the records that one teacher of infants made when talking to groups of about six children:

> *Chalk is used for writing on a blackboard because it's soft. It's white. If it was black you couldn't see it on the board.*

> *Glass is used for windows because you can see through it. It lets the light in. It keeps out the rain. The sun comes through in summer to make us warm.*

> *Wood is used for doors, cupboards, shelves, tables and chairs because it's hard. It doesn't go soggy. It keeps out the rain. It doesn't tear or bend. It's firm and strong.*

Two characteristics of the chalk were mentioned as making it suitable for its job, its softness and its colour. Three properties of glass were mentioned: transparency, impermeability to water and transmission of heat. The latter is a function of the thickness as well as the nature of the material and points to the occasional difficulty in distinguishing a general property of a material from the behaviour of particular objects made of that material. Thus, glass windows do let heat in (and out) fairly well but that is in part attributable to the thinness of the material used. Glass is not a particularly good conductor of heat and if the prime concern was to let heat in, it would not be chosen. Nevertheless, statements about letting heat in reflect some sophistication in the thinking of such young children. Similarly, the comparatively large number of properties attributed to wood indicated progress in both awareness of those properties and in linking them to how the material has been used. At the same time, they provide points for further discussion. It is not immediately clear, for instance, what has been meant by 'hard', 'firm' or 'strong'. Another possible point for development is the idea of wood not going soggy and keeping out the rain. While wooden doors, for instance, do not go soggy in the way paper ones might, they do rot in the long term. Wooden items are normally treated in some way to lengthen the time it takes for the elements to have this effect.

Children's comments occasionally provided scope for intervention of a practical nature. For instance, one group produced the following thoughts about why clay pots were suitable for holding plants:

> *It will easily break if you drop it but we use it to stop the shelf or table getting all wet from the plant. It keeps the water in.*

The disadvantage of the material, its brittleness, was contrasted with the assumed advantage of non-permeability. This provided a good opportunity for testing out the idea that this material did not let water through. A further refinement would have been to compare the porous nature of this material with others such as the plastic also used for plant pots.

Another teacher also found it useful to focus on items for which alternative materials are normally available. Children in that class compared plastic and paper straws. They tested them by using them for drinking and tried to describe the advantages and disadvantages of each material. Each child was able to come to a conclusion as to which was better. They did not all agree because one criterion may have weighed more heavily for one child than another. Nevertheless, comparison of alternative materials, if possible through direct testing, seems a useful extension to work which focuses on the suitability of typical, individual materials.

In fact, children often made comparisons with another material when asked to say why a particular material had a particular use. In this case, however, they contrasted a suitable material with one which they deemed unsuitable. Here are a number of examples of such comparisons:

> *Wood is used to make wardrobes, not glass because it would break.*

> *Sinks are made of pot, not paper because it would tear. Paper would not hold water.*

> *Wool is used to make clothes. It keeps us warm. If sheep had paper coats, they would be cold.*

The statements about the 'unsuitable' materials are sometimes ideas which could be tested. Can paper hold water, for example, and would something be cold with a paper coat?

In other cases, children had ideas about what materials do which were difficult to respond to in that they could not be put to any simple test. Consider, for example, what one child wrote about glass (Fig. 4.15).

Figure 4.15

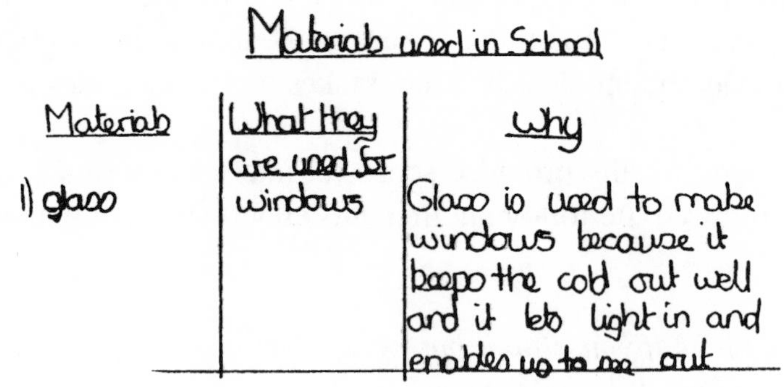

Materials used in School

Materials	What they are used for	Why
1) glass	windows	Glass is used to make windows because it keeps the cold out well and it lets light in and enables us to see out

It is difficult to know how to interpret the expression 'keep the cold out'. It and the expression 'don't let the cold in' may be taken to imply that cold is an entity which can move - from a cold place to a warmer one. Maybe, however, 'cold' is merely a way of referring to cold draughts or wind.

Occasionally, a child seemed unable to give a reason why a particular material served its purpose other than affirming that this was the way things are. This example was taken from the work of a lower junior child (Fig. 4.16).

Figure 4.16

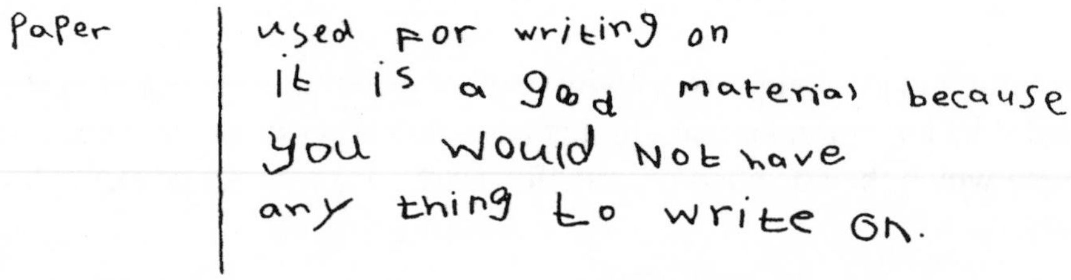

Paper   used for writing on
It is a good material because
you would not have
any thing to write on.

It is as if the paper so perfectly fits its use that no alternative can be considered. Thus, the properties of paper that make it good for writing on were not apparent to the child. Perhaps some development of ideas might take place if the child were asked to think about why another material, say wood or plastic, is not used for writing on.

Some children occasionally gave the function of an object instead of a reason for choosing a particular material. The following upper junior child has often confused material and object (Fig. 4.17).

Figure 4.17

Material Floor tiles	Use walking on.	What make it such a good material. Flat, straight, tessalate, strong, hard wearing, fires easily wiped, hard to crack, smooth.
P.E. mats	binding apperatus	cushioned, spongy, gripping, flexible
plug socket	electric supply.	protecting, adaptable, (can be used for numerous applingsecas), solid, firm, rigid, opeque.
Flourescent lights	to illuminate.	hard to break, bright, long lasting, adaptable, covered, semi-transparent, rigid, & economical.
paint	decoration	blendable, coloured, hard to remove, long lasting, different consistencis, spreadable, washable, easy to use,

Tiles, P.E. mats, plug sockets and fluorescent lights are all objects rather than materials. Some of the stated reasons are equally applicable to both material and object. Both the tile and the material of which it is made could be said to be 'hard wearing' and 'hard to crack'. However, tessellation, for example, is a property of the tiles and their shapes rather than the material. Nevertheless, the range and number of characteristics given indicate well-developed ideas about how properties suit uses.

4.4 Developing Ideas about the Origins and Manufacture of Materials and Changes in Materials

In the exploration phase, children had made drawings to show where they thought certain materials had come from. A necessary prerequisite for understanding the origins of materials is the acceptance that change is possible. Children had been asked about the possibility of changing by heating and of transforming metal during the pre-intervention interview. Hence development of ideas was to be encouraged in both these aspects, that of the origins of materials and that of change in materials.

Activities to develop each aspect evolved from the pre-intervention meeting. Activity 3 'Where does it come from?' pertains to the first aspect, that of origins, while Activity 4b 'Finding out how materials can be changed' pertains to the second aspect, that of change. Both of these are recorded as part of Appendix VI. This section briefly outlines those activities and then indicates some of what teachers actually did using those activities as guidelines as to the kind of intervention that might take place. Teachers adapted the activities to suit their own classes and, as far as possible, the ideas children had expressed. Moreover, the work on the origins of materials required embedding, where possible, in a local context.

4.4.1 Origins and manufacture of materials

Teachers encouraged children to trace back materials to their origins. A number of materials were suggested as emanating from local sources. These were pottery, glass, bricks, ceramics, pencils, clothing, metal items, paper, rubber, plastics and food. Where first-hand experience could not be provided, then secondary resources such as information from firms, books and videos might be used. As a preliminary to their search amongst such information, children could discuss one another's ideas about how particular objects were made and where they came from in the first place. Teachers might also guide children towards considering the various processes of change from raw materials to finished product. These changes might be shown as a flow chart.

The teacher of a group of infants arranged a visit to a local herd of goats. They saw all the processes involved in the production of goats' milk. This included milking, bottling, cooling, storing and packaging. That same group could also remember a visit to a bakery in which they had seen flour turned into bread. Figures 4.18 and 4.19 are drawings made by a couple of the infants. Annotation of what they had said has been added to the drawings. The children have shown some of the processes involved in bread and milk production. They have also added to the sequence other features from their visits that had particularly caught their attention, the eating of the bread and the feeding of a baby goat.

Figure 4.18

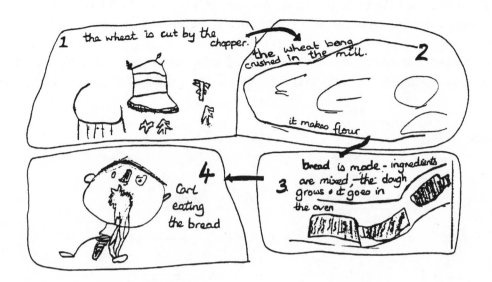

Figure 4.19

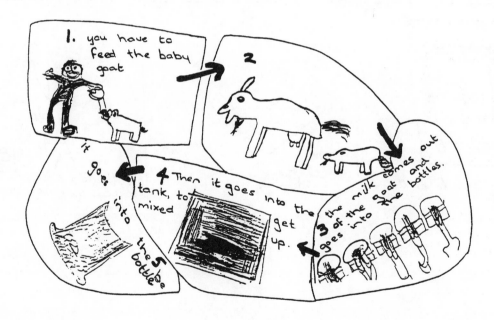

The infants together with junior aged children from the same school also visited a glass museum. Although the teacher felt that the visit had been 'rather above all the children' it had stimulated ideas about glass. She sought to help the children think about glass production in the following way:

We talked about the process and produced simple plasticine models. These were well made and included details such as the 'window' of the furnace which was red with heat and the glass becoming thinner as it was forced through the roller. Some good, descriptive language resulted during the session.

The older children recorded what they had learned about glass production by drawings. The following two drawings were both given the title, 'How a sheet of glass is made' (Figs. 4.20, 4.21).

Figure 4.20

Figure 4.21

How A Sheet of Glass Is Made

He is puting The glass in to The fire

He is blowing the glass

went she is doing to shape it.

The bulb is Fintsd

Whereas the first drawing (Fig. 4.20) shows the initial production of glass from raw materials, the second (Fig. 4.21) starts with the glass already made. In fact, this child seems to have ignored the title and has drawn what perhaps had made a stronger impression during the visit. The glass has been transformed by heat to give a different shape. The bulb-shaped glass becomes an electric bulb although the intervening processes have not been shown.

Some young children made short written accounts about the manufacture of materials to accompany paintings they had also made. Here is one example (Fig. 4.22).

Figure 4.22

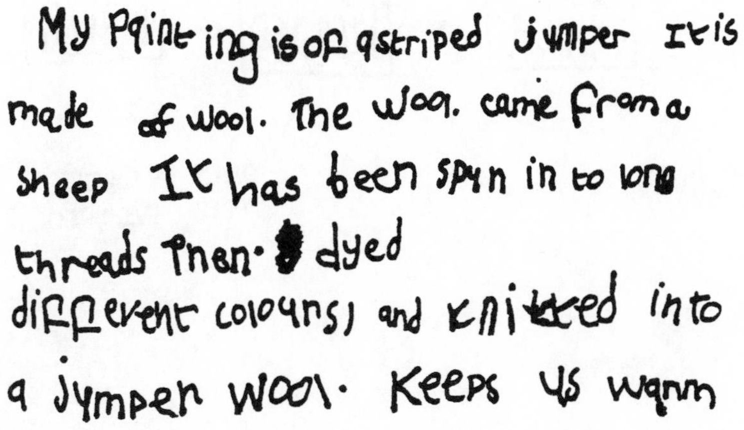

My Painting is of a striped jumper It is made of wool. The wool came from a sheep It has been spun in to long threads then dyed different colours) and knitted into a jumper wool. Keeps us warm

88

Other children, both infants and juniors, produced a series of pictures to represent the sequence of processes involved in manufacture. In the following example (Fig. 4.23) from an infant class, arrows have been used to show the sequence. The child has indicated a number of stages by which a cow has been transformed into a pair of shoes. In one case, a product, shaped pieces of leather, has been labelled as a process, cutting.

Figure 4.23

A lack of differentiation between process and product when representing manufacture by flow charts also occurred for upper juniors. Children of this age group generally show a greater awareness of the number and nature of processes involved in manufacture. The first example (Fig. 4.24) has only words in the boxes while the second (Fig. 4.25) also has pictures.

Figure 4.24

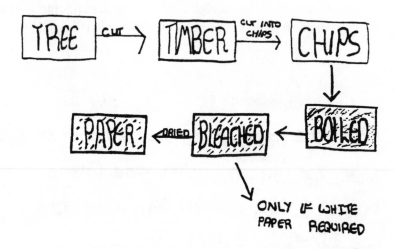

Figure 4.25

In the final two examples, (Figs. 4.26 and 4.27) descriptions of both process and product have been combined in a sequence of pictures with writing. They show how children had been able to glean information from various sources and represent it in their own terms.

Figure 4.26

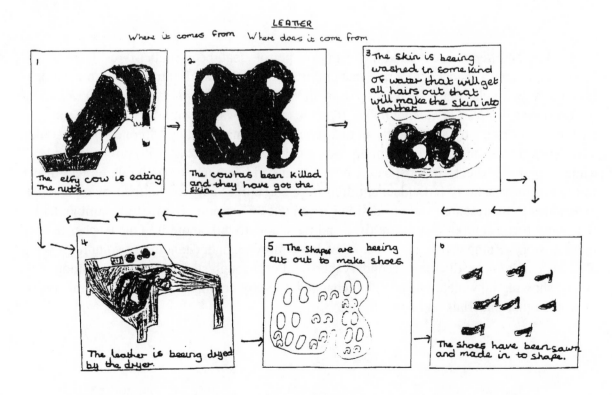

Figure 4.27

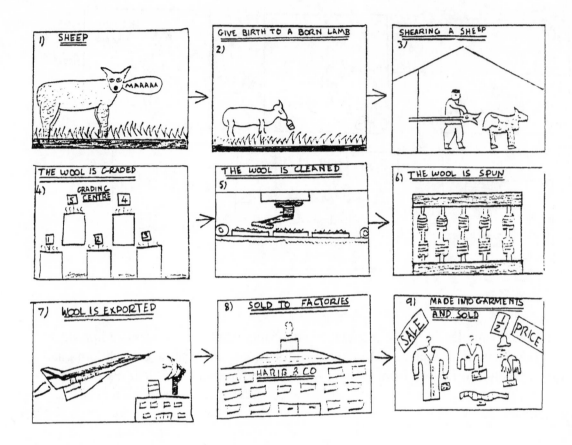

Some older children also explored the origins of petrol, coal, beer, rubber tyres, leather jackets and paper. Teachers reported that children found this work intrinsically interesting. Looking at particular items such as a pair of shoes quickly *stimulated a search for details of the processes involved in making them.* In some cases, it was possible to trace materials back to their origins by a visit to the actual processes of manufacture. In one school, some-one visited a class and showed them how she made jumpers, starting with raw wool and using a spinning wheel. In most cases, however, it had been necessary to rely on less direct sources of information. Individual children were sometimes able to add their own personal experience to the research and discussion carried out by a group. Some, for example, had seen wool being spun or clay being fired and were able to talk about it. One teacher used a familiar story (Rumpelstiltskin) to help develop children's understanding of spinning. What all these experiences, direct or secondary, personal or through stories, have in common is the aim of showing how materials could be changed, thereby helping children to become aware of where other materials had come from.

4.4.2   Changes in materials

Change sometimes results in a total transformation of a material so that the product is completely different from the original material. Glass looks very different from the sand from which it is made. In other cases, the resemblance between original material and product is quite strong. Raw wool and a knitted jumper have similar textures.

Some suggestions about the possible changes that might be considered resulted from the pre-intervention meeting (Activity 4b Appendix VI). These suggestions included changes brought about by heating or cooling, changes brought about by mixing or dissolving, and changes which involved physical treatments of a material such as pressing, rolling, bending and twisting. What follows is an account of some of the changes that children actually investigated.

One class of infants looked at the changes to clay as part of an attempt to help them understand the origins of ceramic material. Here is part of an account of this work by the teacher; the children had made some clay models:

> *When the children handled clay, the general descriptions were 'gungy' 'sticky' 'squashy' 'cold'. I asked what happens to the wet clay to make it like this (a plate). We broke the plate and looked at the cross-section. 'It's painted on the outside.' 'It's hard and it's dried out.' We agreed to dry our models out and then paint them. They were compared to see the drying out process. A three day old model and a day old model were plainly different colours and there was a noticeably colder feel to the fresher one.*

These children experienced two kinds of change. Firstly, they saw that the shape of clay had been changed when squashing it to make their models. Through experiencing a relatively malleable material like clay they might at a later stage come to appreciate that apparently hard materials like metal can also be shaped in a similar way under certain conditions. Secondly, they had also experienced the more permanent change that occurred as clay dries out. By looking at a broken plate, the teacher had helped the children to make a link between this change and the origin of the plate.

A group of junior aged children investigated the changes that could be brought about in milk by adding vegetarian rennet to it. Figure 4.28 shows an account compiled by the whole class. The children made useful comparisons between the properties of the different materials involved, the initial substance, milk, the intermediate liquid and the two substances they had separated with a muslin cloth. They have also traced back the cheese to beyond its origins as milk with an indication of the cow's ability to change grass.

Figure 4.28

Cheese

1. Grass was growing in the field.
2. The cow ate the grass.
3. The cow was milked.
4. The milk was poured into a bucket and a drop of vegetarian rennet was added. It was a warm and creamy white liquid.
5. This was left in a warm place for two hours with a cloth on top. It looked like trifle - yellow around the sides and frothy in the middle.
6. The mixture was scooped into a muslin cloth. It was creamy and like yoghurt or soft blancmange and a yellow juice was left in the bucket
7. We left it all afternoon for the juice to drip out. We added a little salt. When we ate it. it was pretty solid and creamy.

A group of infants made suggestions about how they could change a number of materials in front of them. These included:

plasticine	-	*by pulling and pushing it with your hands*
aluminium	-	*by bending it*
cotton wool	-	*pull it apart, break it*
foam rubber	-	*twisting it*
dried peas	-	*put in water*
coffee granules	-	*squeeze into powder, add water and it disappears*
tile	-	*break it, use a hammer*
jelly	-	*cut it, put hot water on it, it melts and then goes solid again*
rubber	-	*bending it*

They tried some of the changes. For example, they had said that plastic would break. They attempted to break it but only squashed it. It is interesting that, in each case, the material seemed to evoke a specific change. This is not surprising in the case of the food items, where a particular change is normally experienced. Similarly, perhaps, children might be limited in what they could do to a piece of aluminium. The cotton wool, however, might easily be treated in other ways. It might be squeezed or dyed or rolled. However, it seems that these children are used to doing one particular thing with each material.

Several classes looked at changes brought about during food processing. Since these often involve a change of state, teachers also took this as an opportunity to develop ideas about solids, liquids and gases. Some of those activities are mentioned in section 4.3 which reports on intervention in that area. They include beating egg whites and making scrambled eggs. Also mentioned there is the search by some upper junior children for real-life instances of the effect of heat. One further example is mentioned here.

A number of children in different classes looked at the preparation of jelly. Here is one account, by a lower junior child, of the changes that took place (Fig. 4.29).

Figure 4.29

> Jelly
> 1. When the jelly was unwrapped it was stretchyand bendy hard to break
> 2. when boiling water was add to the Jug it started to disintegrate the cubes like Ice bregs melt ing
> 3. When we ate it it was lovely. it wds Soft and wobbley because weleft it over night

The differing properties of the jelly cubes and set jelly have been referred to. Similarly, some infants noted differences in appearance and taste between jelly cubes and set jelly but they also pointed out that, on cooling, the jelly returned to 'like it was before'. That is, it had regained its original state. It is also interesting that of two terms applicable to the change on heating, some children used the word 'dissolved' whereas others chose 'melting'.

A final word of caution needs to be added about the impression that this chapter gives. A rich and varied intervention has been portrayed. Teachers worked in different ways to encourage children to develop their ideas. However, the actual intervention undertaken by any one teacher or experienced by any one child was clearly much more limited than the 'overall picture' that has been presented. The time period for intervention was five weeks and a few teachers had an opportunity for only four or five 'science' sessions. Where children became absorbed in one particular area for development, other areas necessarily received less attention and occasionally were not addressed at all. In particular, teachers felt that they had not been able to expose children to as full a range of practical investigations of the properties of materials as they would have wished.

5. *EFFECTS OF INTERVENTION: CHANGES IN CHILDREN'S IDEAS*

5.0   Introduction

The classroom activities, some of which are described in the previous chapter, took place over a period of four to five weeks within a very tight schedule. Although there was considerable communication between teachers and the research team which permitted a certain amount of exchange of information, it must be stressed that children's initial interview responses and classwork had not been summarised, analysed and interpreted when the class work was instigated. The teachers had met as a group; the prevailing ideas which had been elicited from children were identified and discussed, together with a consideration of what general strategies and specific activities might be appropriately adopted. Though more focused than an informal exchange of ideas, the outcomes of this kind of meeting could not be described as the specification of a controlled treatment programme within a pre- and post-treatment research design. What evolved from the teachers' discussions was much more akin to good practice in planning with the benefit of exchanges with peers from other schools.

The general principle was to start with the ideas which children had expressed and to support development in the direction of scientific thinking from that point. This involved teachers in moving from a more passive, listening stance in which the encouragement of the expression of ideas, all of which would be treated as provisional, was the paramount concern, to a subtly more active, challenging and questioning style which encouraged children to seek evidence and offer justifications for their ideas. While all teachers would be attempting to direct children's learning experiences towards conventional scientific understanding, there was no attempt by the project to constrain the particular pathways and experiences which might unfold within individual classrooms. Indeed, it was anticipated that, within the general mode of operation suggested by the project, teachers would discover or invent novel ways of helping children to make progress in their understanding of various aspects of materials. This was confirmed in practice.

The structure of this chapter parallels that of Chapter Three; the earlier chapter describes and illustrates the ideas which children expressed prior to the period of classroom activities. This chapter documents any shifts which became apparent in children's thinking following that period. In view of the comments above which make clear that the classroom activities were not intended to be tightly specified, it is not possible to attribute particular outcomes to particular experiences. However, there are indications of particular areas of thinking in which it seems that shifts can be achieved, and from within the classroom repertoires to which children were exposed, some productive strategies can be identified.

5.1 Descriptions and Properties of Materials

5.1.1 Classification of materials

The initial classification activity was separated from a repeat of the exercise by the three week long classroom intervention period. The classroom activities focused on the properties of materials through various investigations, rather than properties *per se*. Detailed initial and post-intervention data were available for 52 of the children reported in Table 3.2, page 22. The number of set labels common to each individual's pre- and post-intervention classification is shown in Table 5.1 below.

Table 5.1 Incidence of individuals' use of set labels common to initial and post-intervention classification (n=52)

Percentage of sample	12	21	27	13	13	8	4	-	-	2
Number of children	6	11	14	7	7	4	2	-	-	1
	0	1	2	3	4	5	6	7	8	9

Number of classifications in common

Only six children (12 per cent) changed their set names so that there was nothing in common between the two responses; at the other extreme, one child matched each of her nine set labels perfectly despite the three-week interval. About half the children used fewer sets for the later classification, 37 per cent used the same number and 15 per cent used more groups in the final classification.

The incidence of the most commonly occurring set names showed minimal change. In order of decreasing frequency of occurrence these were, post-intervention: 'food', 10 per cent; 'metals', nine per cent; 'natural/from nature', five per cent; 'building materials', five per cent; and 'materials', three per cent.

There is evidence to suggest that children tended to be thinking in compositional terms to a greater extent after the three-week intervention period than they had been before and that this shift was at the expense of functional classifications.

5.1.2 Judgements about the properties of materials

It was indicated in Chapter Three that many children made very little distinction between the properties of hardness and strength. In the context of the value which the project placed on active exploration and testing of ideas, properties of materials would be seen as particularly amenable to active investigation in the classroom. In other words, these properties are capable of being investigated. A necessary first step would have been for children to operationalise those definitions which they currently held.

A significant difference in the presentation of the task, post-intervention, was that the materials to be considered were those identified by the interviewer rather than those selected by the interviewee. The comparisons which children were asked to consider were which of two metal strips was harder, and which of cotton and woollen threads was stronger.

Deciding which of two metals might be harder

Several responses had been prominent pre-intervention: malleability, resistance to intrusion, ability to resist impact and structural rigidity. The last of these was closest to the dominant idea post-intervention - a consideration of whether or not the metal would bend under a load.

Table 5.2   Criteria used to judge the 'harder' of two pieces of metal

	Post-intervention		
	Infants **n=23**	**Lower Juniors** **n=23**	**Upper Juniors** **n=21**
Observational criteria:			
Size, thickness	–	13 (3)	5 (1)
Feel, touch, texture	4 (1)	–	–
Other	–	17 (4)	–
Don't know	–	–	14 (3)
Criteria implying test:			
Makes a sound on impact	4 (1)	–	–
Impresses, is malleable (dents, or can be squeezed)	4 (1)	–	–
Resists cutting, abrasion	9 (2)	4 (1)	–
Breaks/shatters on impact	13 (3)	17 (4)	14 (3)
Bends under load	65 (15)	40 (9)	67 (14)
Other	–	9 (2)	–

It is not possible to say to what extent this expression was attributable to the particular nature and shape of the metal items presented, or the extent to which classroom experiences had moved children towards this particular understanding. For example, observational responses referring to heaviness were in evidence in 15 per cent of responses, pre-intervention. In the second round of interviews, there were no references to heaviness. One plausible explanation is that the metal strips were not, relatively speaking, 'heavy', nor was the difference in their weights (or masses) substantial. This is to suggest that children may be susceptible to very local and specific cues about properties, based on the overall configuration of the specific object in view.

One point which does seem to be unequivocal is the shift of the sample, the infants in particular, towards an operational and testable definition of the property under discussion. Only one infant, seven lower juniors and one upper junior used observational criteria in expressing their judgement as to which of the two metals might be harder. This does not necessarily imply that children ended up with a better understanding of the property which they were trying to test. This depends not only on the accuracy of their investigation but also on the accuracy of their operational definition. For example, in their struggle to tackle the concept of 'harder', a number of children were inadvertently testing bending. That is, they were testing 'Which of the two metals is harder to bend until it breaks?' In effect, they were victims of the common metaphorical use of 'hard' to mean 'difficult'. Their operational understanding of the property of hardness might not have advanced; they will have explored flexibility systematically. Having had the physical experience of grappling with the properties of metals, a teacher-led class discussion about the differences between metaphorical and physical uses of the word 'hard' could have been opportune and meaningful.

Deciding which of cotton and woollen thread might be stronger

Asking children how they would determine the relative strength of cotton and woollen threads gave rise to very few observational responses; a minority referred to the relative sizes of the threads, or their relative weights (see Table 5.3).

> *Wool is fatter and cotton's thin, so wool would be strongest.*   Y3 B L

Table 5.3   Criteria used to judge stronger: cotton and woollen threads

	Post-intervention		
	Infants **n=23**	**Lower Juniors** **n=23**	**Upper Juniors** **n=21**
Observational criteria:			
Size, thickness	9 (2)	13 (3)	–
Heaviness	4 (1)	4 (1)	–
Appearance	– .	4 (1)	–
'Because of what it's made of'	–	4 (1)	–
Criteria implying test:			
Makes a sound	4 (1)	–	–
Resists cutting, abrasion	4 (1)	4 (1)	–
Breaks when pulled	52 (12)	62 (14)	81 (17)
Stretches when pulled	9 (2)	–	5 (1)
Unspecified outcome of pulling	17 (4)	9 (2)	9 (2)
Don't know	–	–	5 (1)

The overwhelming response was to suggest that relative strength would be tested by pulling the threads, with the majority of children specifying the kind of outcome which they would be looking for. It might be tempting to conclude that to test the relative strengths of the

100

threads by pulling was an obvious, common-sense response. A detailed examination of precisely how children described their tests (Table 5.4) reveals that there was a good deal of scientific reasoning involved.

Table 5.4   Nature of test of relative strength of threads

	Post-intervention		
	Infants n=23	Lower Juniors n=23	Upper Juniors n=21
Dependent variable not specified:			
Mention of pulling or how 'hard to break' only	22 (5)	22 (5)	29 (6)
Dependent variable specified:			
First to break	52 (12)	35 (8)	47 (10)
Dependent variable specified and controlled:			
Use of a fixed pull to compare breaking point to break threads	4 (1)	9 (2)	5 (1)
No test by pulling:	18 (4)	30 (7)	5 (1)
Don't know/no response	4 (1)	4 (1)	14 (3)

Most children not only suggested a test; they went on to specify the precise evidence or outcome which they would be looking for upon which to base their decision, i.e., they specified the dependent variable in their investigations. Clearly the test, for most children, was not just a simple everyday matter of pulling the threads between the hands. Most tests involved comparisons of outcomes, and a minority involved something closer to a quantified approach using the gradual addition of weights.

> *Tape the thread to the chair, tie the end in a loop and put a hook on it.*
> *Put a margarine tub on the hook and put weights in until it snaps.*   Y6 G M

SPACE Report   *Materials*

You couldn't just pull one and see if it snapped and then pull the other because you don't know if its the same force. Get something which is strong and tie the string to that and put it high, and get a yoghurt carton and put weights on and see what snaps. I think cotton would be strongest. You could do them one at a time or both together, because you've got the same force with using weights.   Y3 B M

The difference in overall levels of performance before and after the classroom activities can be summarised as follows. The proportion of children implying pre-intervention: the necessity or advantage of an empirical test of 'harder' and 'stronger' was 40 per cent, 26 per cent of infants, 61 per cent of lower juniors and 32 per cent of upper juniors. In response to different items (materials which probably offered a stronger invitation to conduct a test) and following the classroom intervention period, 75 per cent consistently implied the use of tests to decide which of two materials was the harder and stronger, comprising 87 per cent infants, 61 per cent lower juniors and 76 per cent upper juniors.

5.2   Solids, Liquids and Gases

5.2.1   *Assumptions about gases*

The drawings which teachers had asked children to produce during the classwork were a useful source of information about the familiarity of materials in various states. Children were asked to divide a page into three sections and draw whatever materials they thought appropriate under the headings of 'solids', 'liquids' and 'gases' (cf p37). The kinds of examples which children selected were of interest, as were the relative frequencies with which the three states of matter were exemplified. Although all children in any given class might have tackled this task or one very similar, detailed analysis was only undertaken in respect of the responses from the individually interviewed sub-sample. Table 5.5 shows the relative extent to which materials in different states were depicted in those drawings, before and after the intervention period.

Table 5.5   Mean numbers of solids, liquids and gases shown in children's drawings

	Pre-intervention			Post-intervention		
	Inf n=14	**LJ** n=19	**UJ** n=20	**Inf** n=14	**LJ** n=19	**UJ** n=20
Examples of hard strong solids given	0.9	3.7	5.5	6.4	4.2	8.5
Examples of other kinds of solids given	0.4	0.1	0.6	0.8	1.0	3.2
Correct examples of all kinds of solids given	1.3	3.8	6.1	7.1	5.2	11.7
Correct examples of liquid given	1.0	2.4	4.4	4.1	6.0	8.3
Correct examples of gas given	0.6	1.2	3.8	1.0	2.1	4.2

Consistently, fewer gases were indicated in children's drawings than either solids or liquids. (Solids and liquids will be discussed further in the next section.) Following intervention, there was an increase in the number of references to gases and the strong age trend remained in evidence. Gases are not as familiar to children as are materials in their solid or liquid states, but seem to become more so as children increase in age and experience. Since the gases with which children come into contact most frequently (those in the air) are transparent and odourless, this lack of familiarity is not too surprising. (Intervention work on gases with infants was very limited, except through association with carbonated drinks.)

The kinds of gases (and associated concepts) which children showed in their drawings are summarised in Table 5.6.

Table 5.6   Response categories: drawings of gases

	Pre-intervention			Post-intervention		
	Inf n=14	**LJ** n=19	**UJ** n=20	**Inf** n=14	**LJ** n=19	**UJ** n=20
Categories:						
Word 'trigger'	50 (7)	42 (8)	50 (10)	21 (4)	37 (7)	40 (8)
Air	–	–	35 (7)	21 (3)	37 (7)	50 (10)
Smoke/fumes	7 (1)	5 (1)	25 (5)	–	16 (3)	25 (5)
Association with liquid	–	26 (5)	40 (8)	29 (4)	16 (3)	25 (5)
Named gases	–	5 (1)	60 (12)	–	16 3	65 (13)
Association with fire/heat	64 (9)	26 (5)	50 (10)	21 (3)	37 (7)	15 (3)
Association with danger	7 (1)	–	5 (1)	–	–	–
Non-gas	14 (2)	42 (8)	20 (4)	7 (1)	16 (3)	30 (6)

It is immediately apparent that many children's selections of gases to include in their drawings were prompted by a word 'trigger'; that is, the name of the material in conventional usage included the word 'gas': gas cooker, gas heater or, more rarely, natural gas, camping gas, etc. The proportion of this kind of response can be seen to have declined in the follow-up interviews. The decline in responses relating gas to fire and heat is also connected with this word 'trigger' type of response, since the connection with heat was often in the form of gas fire or gas cooker.

The major increase was in the number or responses which made a connection between gas and air; children had obviously acquired the knowledge that air is a gas during the intervention period. It was not uncommon, pre-intervention, to find that children did not think of air as being a gas. Following the discussion about the 'empty' container reported in section 3.2.1, children were asked some more direct questions about air, including whether they con-

104

sidered air to be a gas. The following responses illustrate the distinction which children sometimes made between air, which was perceived as being benign, and 'gas' which often held sinister connotations of danger:

Q.	What is gas?
R.	*Gas is a fire.*
Q.	Is air a gas?
R	*No. Air isn't a gas.*	Y3 B M

R.	*Gas can be dangerous.*
Q.	Is air a gas?
R.	*No.*	Y3 G M

Q.	Is air a gas?
R.	*Air isn't a gas cos gas could kill you, but air can't.*	Y3 B M

The following dialogue provides an example of a child who has ideas about gases being dangerous, but with the realisation that there are also useful gases.

Q.	What's your idea about gases? What are gases like?
R.	*They are like that we could either breathe in, like oxygen, or they could be a poisonous kind of gas I can't think of the name of.*
Q.	Can you say anything else about gases? What are they like?
R.	*You can't see them.*
Q.	Can you never see gases?
R.	*You can see the sparks of a fizzy drink when they all bubble up.*	Y6 G M

Table 5.7 shows the proportions of children who gave no correct examples of a gas or gave word-triggered examples only.

**Table 5.7	Children giving no correct example of 'gas' or giving only examples triggered by the word 'gas'**

	Pre-intervention			Post-intervention		
	Inf n=14	LJ n=19	UJ n=20	Inf n=14	LJ n=19	UJ n=20
Gave no examples or examples given are incorrect	36 (5)	16 (3)	–	36 (5)	26 (5)	5 (1)
Gave word-'triggered' examples only	64 (9)	47 (9)	20 (5)	14 (2)	5 (1)	–

Table 5.8 summarises the properties of gases offered by children when asked directly. There is a clear age trend in evidence, indicating that the older children were able to suggest more properties than were the younger ones.

Table 5.8   Defining attributes of 'gas' offered

		Post-intervention		
		Infants **n=23**	**Lower Juniors** **n=23**	**Upper Juniors** **n=22**
Visibility:	Invisible	–	13 (3)	23 (5)
	May be visible	4 (1)	13 (3)	41 (9)
Smell:	Has a smell	4 (1)	9 (2)	5 (1)
	May have a smell	4 (1)	4 (1)	14 (3)
Audibility:	May fizz, hiss or be heard	9 (2)	9 (2)	–
Relates to other perceptible forms:				
	Spray, drops, steam, cloudy	–	13 (3)	14 (3)
	Container determines shape	–	9 (2)	–
	Runs through fingers ('unholdable')	–	–	5 (1)
	Relates to smoke	9 (2)	4 (1)	18 (4)
Flammability:				
	Used in cooker	–	13 (3)	5 (1)
	Used in (gas) fire	4 (1)	9 (2)	14 (3)
	Burns	9 (2)	9 (2)	5 (1)
Other responses				
	Relates to air and/or breathing	13 (3)	35 (8)	46 (10)
	It's light/floats in air	13 (3)	13 (3)	18 (4)
	Named gas	–	13 (3)	32 (7)
Mean number of properties of gas offered per child		0.7	1.7	2.4

A number of children can be seen to be grasping for tangible properties of gases and relating these to the more familiar perceptible properties of spray and smoke.

5.2.2   *Identification of Materials as Solid or Liquid*

Solids

Table 5.5 above summarises the proportions of solids, liquids and gases which children included in their drawings. Solids were clearly the most familiar materials in children's perceptions. It is also evident that solids which can be identified as having the properties of hardness and strength were those which predominated. It may be inferred that these properties comprise the core attributes which define many children's construct of the concept of 'solid'. Following the period of classroom intervention there was an apparent broadening of the definition to include other kinds of solids, particularly amongst the juniors. The fine detail of this conceptual broadening is summarised in Table 5.9.

Table 5.9   Response categories: drawings of solids

	Pre-intervention			Post-intervention		
	Inf n=14	LJ n=19	UJ n=20	Inf n=14	LJ n=19	UJ n=20
Categories:						
Hard strong solids	71 (10)	95 (18)	95 (19)	93 (13)	100 (19)	95 (19)
Soft, malleable solids	– 	11 (2)	20 (4)	50 (7)	47 (9)	75 (15)
Powdery solids	7 (1)	–	5 (1)	–	26 (5)	15 (3)
Geometrical shapes	7 (1)	5 (1)	5 (1)	–	–	15 (3)
Non-solids given	14 (2)	–	–	–	–	5 (1)
Include composite objects	7 (1)	47 (9)	30 (6)	57 (8)	26 (5)	20 (4)

Almost all children included at least one example of a hard, strong solid in their drawings, post-intervention. The enormous shift is in the inclusion of soft, malleable materials within the category, and to a lesser extent, the allowance of powdery solids. Composite objects, those made of more than one material, were also more frequently in evidence amongst the infants: a building, a person, a wall, etc. In some cases, there was a manifest understanding that such a composite object could be a member of two physical state sets simultaneously.

It remains the case, despite the shifts described, that half the infants and lower juniors and a quarter of the upper juniors did not include any malleable solids in their drawings. Powders remained even more problematic as instances of solids for all age groups. Figure 5.1 reproduces the drawing made by the same child whose pre-intervention sets of materials are reproduced as Figure 3.4 on page 39.

Figure 5.1   Drawing of solids, liquids and gases

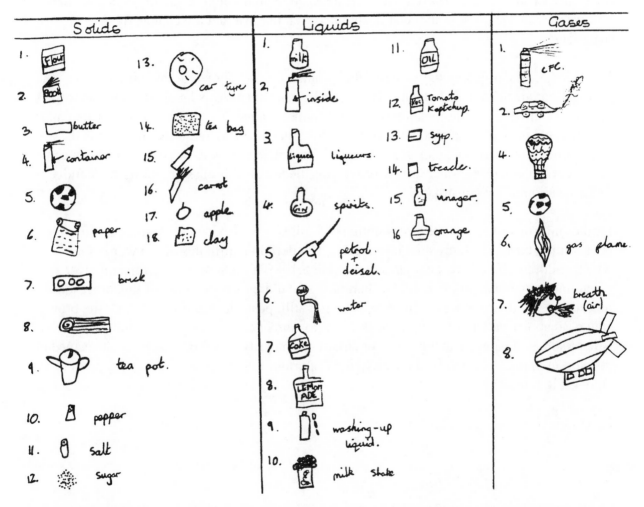

This single drawing very effectively illustrates some of the shifts which were evident in the sample as a whole. From 10 solids, seven liquids and three gases, there has been an increase in the numbers of materials represented to 18 solids, 16 liquids and eight gases. The overall number of items identified is higher than average but in line with the general trend. Amongst the solids drawn in Figure 5.1 are some types which were totally absent in this child's initial response to the task, several weeks earlier. For example, there is flour, sugar, salt, pepper, and a tea bag, all of these being examples of solids in a granular or powdery form. There are also hollow solids - the ball, tyre, tea pot and container. There are flimsy solids such as the paper, as well as vegetable material - the carrot and the apple. Butter, a malleable solid, is included. The number of objects has increased, but this is not because there are more of the same; it is associated with a subtle redefinition of the entry qualifications for set membership beyond hard and strong structurally rigid objects alone.

Of the sixteen liquids, eleven are drinkable - about the same proportion as previously. Interesting additions are the more viscous liquids: milk-shake, tomato ketchup, syrup and treacle. Liquid number two would be a potentially interesting starting point for a discussion with the child, the liquid in a spray canister having been indicated under the set label of liquid, and the expelled material indicated under the set label of gas. This observation of the difference in state of the material when inside as compared with outside the container may motivate a consideration of the relationship between the two and the change of state of materials.

Invisible gases, defined by the shape of their containers, are shown in the third column - the football, hot air balloon and airship. The topically relevant CFC is labelled emerging from the spray canister. The designation of the gas flame as a gas invites further discussion, as does the inclusion of motor exhaust. Is the visible flame considered to be the gas? Is the exhaust thought of as smoke or gas? The set definition has shifted, but there will be uncertainties at the boundary which would prove a fertile area for discussion during the child's reorganisation of the construct.

Further information about children's notions of solids was provided more directly through questioning during the individual interviews. The defining attributes of the word 'solid' which children offered are summarised in Table 5.10. This summary confirms the overwhelming importance of the notion of hardness in children's conception of the term 'solid'; indeed, the incidence of this attribute was higher still, post-intervention. Despite the provision of a greater variety of exemplars, the 'set' towards hardness as a determining characteristic of solids remains. Having shape or structural integrity was a property which seemed to develop in importance, while in contrast, the importance of 'heaviness' declined, particularly amongst the infants.

Table 5.10   Defining attributes of 'solid' offered

	Pre-intervention			Post-intervention		
	Inf n=23	LJ n=23	UJ n=22	Inf n=23	LJ n=23	UJ n=22
Being hard or rigid	9 (2)	52 (12)	55 (12)	65 (12)	87 (20)	64 (14)
Having shape (stays still or together)	9 (2)	4 (1)	27 (6)	35 (8)	30 (7)	23 (5)
Rattles, makes a noise	4 (1)	13 (3)	9 (2)	4 (1)	4 (1)	–
Heaviness	26 (6)	22 (5)	14 (3)	–	22 (5)	9 (2)
Solid 'feel'	4 (1)	4 (1)	–	4 (1)	–	14 (3)
Other	9 (2)	9 (2)	23 (5)	9 (2)	26 (6)	68 (15)

Twenty-nine per cent of the sample (five infants, four lower juniors, 10 upper juniors) explicitly stated that the 'powder' form of the talcum excluded it from being labelled 'solid'. 'Powder' was treated as a separate state. This reduced to just seven children (10 per cent of the sample) post-intervention. Similarly, nine per cent referred to the cotton wool as 'material', rather than 'solid' or 'liquid'. A third of the infants excluded cotton wool from the solid category on the grounds that it was 'soft'. This response had almost ceased to be encountered post-intervention, being offered by only one infant and five juniors.

Liquids

The kinds of things which children saw relevant to include as liquids are summarised in Table 5.11.

Table 5.11   Response categories: drawings of liquids

	Pre-intervention			Post-intervention		
	Inf n=14	LJ n=19	UJ n=20	Inf n=14	LJ n=19	UJ n=20
Categories:						
Word 'trigger'	64 (9)	68 (13)	50 (10)	43 (6)	47 (9)	70 (14)
Water	–	26 (5)	70 (14)	71 (10)	68 (13)	80 (16)
Drinks	–	42 (8)	70 (14)	71 (10)	89 (17)	85 (17)
Other aqueous liquids	14 (2)	37 (7)	40 (8)	43 (6)	53 (10)	75 (15)
Very viscous liquids	14 (2)	11 (2)	15 (3)	29 (4)	16 (3)	45 (9)
Coloured liquids	14 (2)	68 (13)	90 (18)	71 (10)	84 (16)	90 (18)
Non-aqueous liquids	–	26 (5)	60 (12)	21 (3)	32 (6)	50 (10)
Non-liquids	–	11 (2)	20 (4)	21 (3)	–	10 (2)

As with their drawings of gases, there were significant numbers of illustrations of liquids which seem to have been stimulated by the everyday reference to the item including the word 'liquid', which, it is assumed, acted as a 'trigger'. Examples include 'Fairy Liquid', other washing-up 'liquids', etc. Sixty-five per cent of the infant sub-sample offered examples of liquids of this nature, pre-intervention, while a further 21 per cent of the same group offered either inaccurate examples or no example at all. These word-'triggered' responses decreased amongst the infants and lower juniors, but increased amongst the upper juniors.

Table 5.5 shows that the examples of liquids offered quadrupled for infants, and roughly doubled for the two older age groups. It is possible that their teachers directed some children away from word-'triggered' responses.

The infants seem to have had their eyes opened to the fact that water and other drinks were also valid examples of liquids which could be included in their drawings. Possibly they ignored the obvious in their search for examples of a more 'scientific' or exotic kind, or those specifically including the label 'liquid'. Another shift which was apparent was the inclusion of very viscous liquids which had previously tended to be excluded from any specified state, tending to be regarded as neither solid nor liquid.

The individual interviews provided the opportunity to ask children about the nature of liquids; Table 5.12 summarises the attributes which recurred in their descriptions.

Table 5.12   Defining attributes of 'liquid' offered

	Pre-intervention			Post-intervention		
	Inf n=23	LJ n=23	UJ n=22	Inf n=23	LJ n=23	UJ n=22
Is water, contains water or looks watery	4 (1)	13 (3)	32 (7)	22 (5)	22 (5)	23 (5)
Goes everywhere, is runny, can be spilled/poured	26 (6)	48 (11)	73 (16)	78 (18)	78 (18)	91 (20)
Can be drunk	-	4 (1)	4 (1)	4 (1)	9 (2)	9 (2)
Has bubbles	4 (1)	4 (1)	5 (1)	4 (1)	9 (2)	5 (1)
Other	17 (4)	26 (6)	18 (4)	4 (1)	35 (8)	32 (7)

The association (or similarity in appearance) with water - being composed of water, containing water or looking like water - was a strong factor, being included within more than 20 per cent of responses, post-intervention.

112

The property which children appreciated in greater numbers during the second round of interviews was the capability of being poured, spilled or being 'runny' which is shared by all liquids.

Solid or liquid?

The fundamental property of physical state of objects prompted a focused investigation of children's use and understanding of the terms 'solid' and 'liquid' during the individual interviews. Five small containers of similar dimensions were presented, each containing a different material. The materials were: a steel rod, 1cm diameter by about 10cm in length; a quantity of talcum powder; vinegar; brown treacle; cotton wool. Children were asked to examine the contents of the containers carefully, without opening them, before being asked 'Which of these are solids?' and 'Which of them are liquids?' This was not an easy decision for many children, nor was it necessarily seen by them as an exhaustive division; many were content to classify some of the materials as neither solid nor liquid. Table 5.13 summarises children's allocation of the materials to the categories 'solid', 'liquid' or 'neither of those'.

The identification of a steel rod and vinegar as solid and liquid respectively seemed unproblematic to the juniors. About two-thirds of the infants made this categorisation accurately initially, but after the intervention, the success rate of the youngest group was equivalent to that of the older children. It would appear that there was little conceptual burden to overcome in adopting the labels 'solid' and 'liquid', at least with the example of the hard, strong solid and the very 'runny' liquid, which are almost archetypal in character. As can be seen from Table 5.13, the other materials posed a tougher challenge to children's existing construct system.

Table 5.13   Identification of materials as solid or liquid

	Pre-intervention			Post-intervention		
	Inf n=23	LJ n=23	UJ n=22	Inf n=23	LJ n=23	UJ n=22
Steel identified as:						
Solid	61 (14)	100 (23)	100 (22)	96 (22)	100 (23)	100 (23)
Liquid	–	–	–	4 (1)	–	–
Neither	30 (7)	–	–	–	–	–
Don't know, etc.	9 (2)	–	–	–	–	–
Vinegar identified as:						
Solid	22 (5)	4 (1)	–	4 (1)	–	–
Liquid	65 (15)	96 (22)	100 (22)	96 (22)	100 (23)	100 (22)
Neither	4 (1)	–	–	–	–	–
Don't know, etc.	9 (2)	–	–	–	–	–
Treacle identified as:						
Solid	17 (4)	17 (4)	18 (4)	4 (1)	13 (3)	9 (2)
Liquid	39 (9)	44 (10)	73 (16)	91 (21)	83 (19)	86 (19)
Neither	35 (8)	39 (9)	9 (2)	4 (1)	4 (1)	5 (1)
Don't know, etc.	9 (2)	–	–	–	–	–
Cotton wool identified as:						
Solid	17 (4)	4 (1)	41 (9)	65 (15)	39 (9)	77 (17)
Liquid	5 (1)	–	–	4 (1)	4 (1)	–
Neither	70 (16)	96 (22)	59 (13)	30 (7)	57 (13)	23 (5)
Don't know, etc.	9 (2)	–	–	–	–	–
Talcum powder identified as:						
Solid	30 (7)	–	23 (5)	44 (10)	30 (7)	50 (11)
Liquid	22 (5)	4 (1)	5 (1)	30 (7)	13 (3)	23 (5)
Neither	39 (9)	96 (22)	73 (16)	26 (6)	57 (13)	27 (6)
Don't know, etc.	9 (2)	–	–	–	–	–

Three-quarters of the upper juniors identified the viscous brown treacle as a liquid, but less than half of the younger age groups did so, about one-third of the younger children describing the treacle as 'neither solid nor liquid'. Once again, after the classroom work, the responses of the younger children were on a par with those of the older ones.

The majority response pre-intervention at all ages in relation to the cotton wool was to suggest that it was neither solid nor liquid, though 41 per cent of upper juniors correctly identified it as a solid. The post-intervention gains in children's acceptance of cotton wool as having the properties of a solid were not as extensive as was the case of their acceptance of a viscous substance as a liquid.

The solid/liquid dichotomy seemed to have even less relevance to children in their classification of talcum powder than it had for cotton wool. Initially, most children described it as being neither solid nor liquid. Although there was a substantial increase in the number of children perceiving a powder as a solid during the second round of interviews, this perception was not nearly as widespread and the shifts were not nearly as extensive as had been the case in respect of treacle and cotton wool.

Table 5.14 shows the number of correct classifications of the five materials which children made.

Table 5.14   Identification of materials as solid or liquid: patterns of responses

	Pre-intervention			Post-intervention		
	Inf n=23	LJ n=23	UJ n=22	Inf n=23	LJ n=23	UJ n=22
All Correct Solid: cotton wool, steel, talc Liquid: treacle, vinegar	13 (3)	–	14 (3)	44 (10)	22 (5)	41 (9)
Four Correct Solid: steel, cotton wool Liquid: treacle, vinegar Other: talc	–	–	14 (3)	4 (1)	9 (2)	18 (4)
Four Correct Solid: steel, talc Liquid: vinegar, treacle Other: cotton wool	4 (1)	–	5 (1)	–	4 (1)	5 (1)
Three Correct Solid: steel Liquid: vinegar, treacle Other: talc, cotton wool	9 (2)	48 (11)	36 (8)	22 (5)	35 (8)	5 (1)
Three Correct Solid: steel, treacle Liquid: vinegar Other: cotton wool, talc	4 (1)	9 (2)	14 (3)	–	9 (2)	–
Two Correct Solid: steel Liquid: vinegar Other: talc, treacle, cotton wool	13 (3)	35 (8)	5 (1)	–	4 (1)	5 (1)

The proportion of children correctly identifying each material as solid or liquid increased from nine per cent to 35 per cent, with the gains seeming to be independent of age. The 'all correct' pattern of response was the most frequently encountered combination; the next most frequent was that in which steel, vinegar and treacle were correctly identified with talc and cotton wool omitted from the classification, either because children did not know what to do with them or because they were regarded as neither solid nor liquid.

There was a significant increase in the number of correct classifications of materials as solid or liquid evident in the post-intervention interviews. This increase in the overall number of correct classifications was seen in each of the three age groups (Wilcoxon Signed Ranks Test; infants, p<.001; lower juniors, p<.001; upper juniors, p<.01).

The pattern of individuals' shifts in their classifications is also of interest. Figures 5.2, 5.3 and 5.4 summarise these shifts for three of the materials: treacle, cotton wool and talcum powder. Each oval represents a response category; numbers in boxes within the oval indicate children who have not shifted their response category over the course of the two interviews. The numbers associated with arrows show the number of children shifting their response from one category to another.

Figure 5.2 shows how children's ideas moved with respect to treacle. All four infants who had initially classified the treacle as a solid, together with seven of the eight who had previously treated it as neither solid nor liquid and two who 'didn't know', accepted treacle as a liquid in the second round of interviews. Only one infant changed the classification from liquid to solid.

The shifts amongst the lower juniors were very similar to those of the infants, six moving from 'neither' and three moving from 'solid' to 'liquid'. There was very little movement in the responses of the upper juniors, most of the children staying with their initial identification of the treacle as a liquid, with just three children, two from 'neither', one from 'solid', joining them.

Figure 5.2   Shifts in individuals' categorisations of treacle

(Infants)

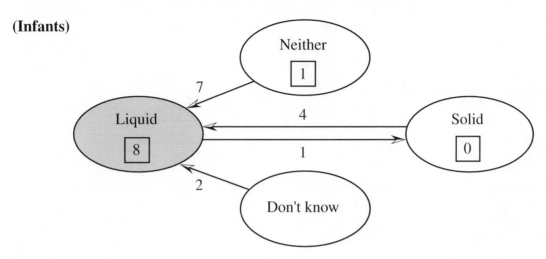

(Lower Juniors)

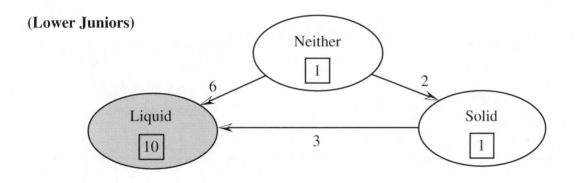

(Upper Juniors)

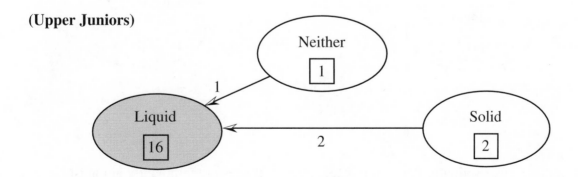

Correct category

Figure 5.3   Shifts in individuals' categorisations of cotton wool

(Infants)

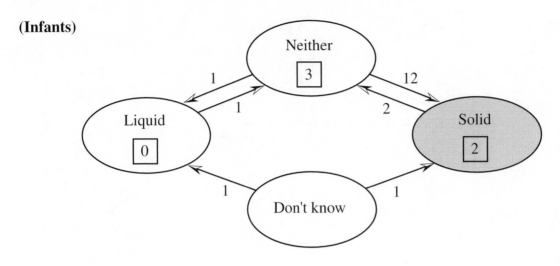

(Lower Juniors)

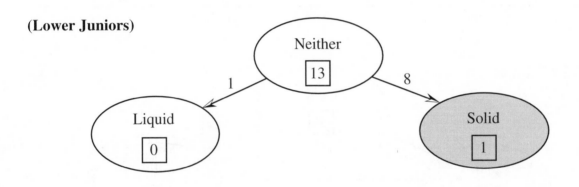

(Upper Juniors)

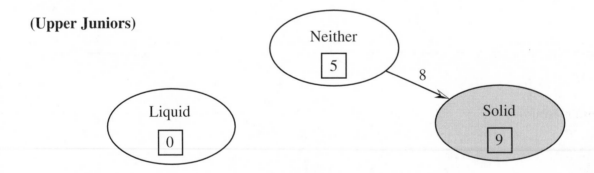

The shifts in categorisation of the cotton wool were at all ages predominantly amongst children who had previously classified it as neither solid nor liquid.

Correct category

Figure 5.4   Shifts in individuals' categorisations of talcum powder

(Infants)

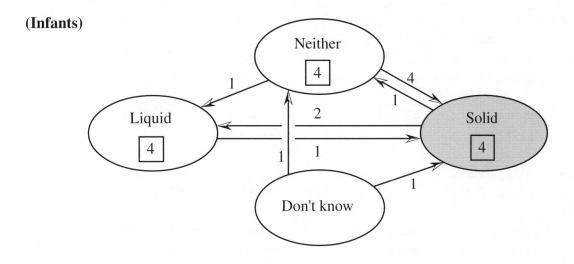

(Lower Juniors)

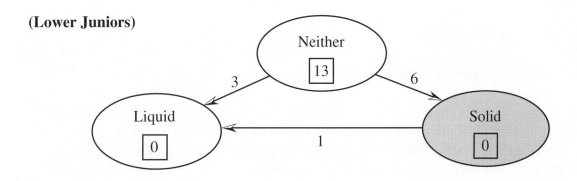

(Upper Juniors)

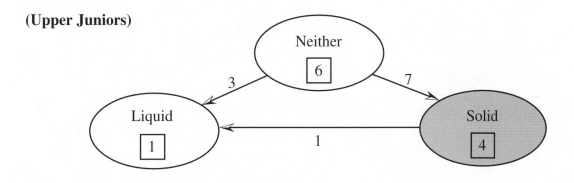

Correct category

120

Figure 5.4 shows children's shifts in ideas about the nature of talcum powder. Amongst the infants, four children consistently regarded it as a liquid in both interviews; two who had previously described it as a solid and one who had categorised it as neither solid nor liquid added to the total in this group. This produced a net increase in the total classifying talcum powder as a liquid. For many children, the attribute of 'runniness' as the defining attribute of a liquid seemed to have been attached to talcum powder, as in the following response:

> *Liquids are all kinds of things that don't stay where you put them. They*
> *just run. If you put them on the table it runs along the table.*   Y1 B H

The mental struggle induced by the task was beautifully articulated by a reception child, who finally decided that the talcum powder should be classified as neither solid nor liquid.

> *It's something like a kind of liquid, but it isn't liquid, it's talcum powder.*
> *It goes fast like vinegar and it's not a solid because you can put your*
> *finger through it. It's a bit solid, but it isn't. You can break it, but it isn't*
> *water. You can put your finger through it, but not like a liquid. A liquid*
> *feels wet and this doesn't.*   R G H

5.3   Uses of Materials

5.3.1   Relating properties of materials to uses

As was noted in section 3.3, the specific object had some bearing on the kinds of uses and properties of materials which children mentioned. This was confirmed during the post-intervention interviews. Children were asked why metal was a good material from which to make nails and inevitably their responses emphasised the capability of metal to be shaped to a sharp point as well as the intrinsic hardness or strength of metal.

Table 5.15   Relating properties of materials to uses: metal

	Post-intervention		
	Infants **n=23**	**Lower Juniors** **n=23**	**Upper Juniors** **n=21**
Functional properties:			
Strong/does not break	57 (13)	61 (14)	64 (14)
Can be sharp (penetrates)	30 (7)	4 (1)	23 (5)
Other	–	9 (2)	9 (2)
Manufacturing properties:			
Can be shaped	9 (2)	9 (2)	–
Aesthetic properties:			
Shiney, pleasing appearance	4 (1)	–	–
Economic properties:			
Overall mean number of properties of metal related to uses	1.0	0.8	1.0

In consequence, the pre- and post-interview questions referring to coins and nails respectively do not offer a useful comparison. (In fact, there was a significant increase in the total number of properties related to use given by the infants; Wilcoxon Signed Ranks Test, $p < .05$.)

A question about the use of wood does permit comparison and performance is summarised in Table 5.16.

Table 5.16   Relating properties of materials to uses: wood

	Pre-Intervention			Post-Intervention		
	Inf n=23	LJ n=23	UJ n=22	Inf n=23	LJ n=23	UJ n=21
Functional properties:						
Strength (structural)	39 (9)	48 (11)	50 (11)	61 (14)	52 (12)	57 (12)
Durability	9 (2)	–	5 (1)	4 (1)	–	5 (1)
Lightness	–	–	9 (2)	–	–	–
Smoothness	13 (3)	13 (3)	14 (3)	4 (1)	9 (2)	14 (3)
Other	4 (1)	17 (4)	36 (8)	22 (5)	35 (8)	29 (6)
Mean number of functional properties mentioned	0.7	0.8	1.1	0.9	1.0	1.0
Manufacturing properties:						
Ease of fixing, jointing (nail, glue)	13 (3)	9 (2)	14 (3)	4 (1)	9 (2)	5 (1)
Can be carved	–	13 (3)	14 (3)	–	13 (3)	19 (4)
Can be cut	–	9 (2)	–	4 (1)	13 (3)	24 (5)
Can be moulded/bent	–	–	5 (1)	– –	–	–
Other	4 (1)	4 (1)	9 (2)	9 (2)	9 (2)	19 (4)
Mean number of manufacturing properties suggested	0.2	0.3	0.4	0.2	0.4	0.7
Aesthetic properties:						
Can be polished, painted, stained etc. (looks nice)	9 (2)	9 (2)	41 (9)	9 (2)	9 (2)	48 (10)
Mean number of aesthetic properties	0.1	0.1	0.4	0.1	0.1	0.5
Economic Properties:						
Cheap, abundant, renewable, etc.	–	4 (1)	5 (1)	–	–	5 (1)
Mean number of aesthetic properties	0	0.1	0.1	0	0	0.1
Overall mean number of properties of wood related to uses	0.9	1.3	2.0	1.2	1.5	2.2

In interpreting Table 5.16 it must be borne in mind that the classification of responses as used to summarise the research was not at that time available to teachers. It is thus a matter of chance whether the particular categories of response - functional, manufacturing, aesthetic and economic - were targeted for attention in the classrooms. It can be assumed in general that teachers attempted through group discussion techniques to encourage a greater awareness of the relationship between the qualities possessed by a material and that material's uses. Table 5.16 reveals that there was an increase in responses describing the functional properties of wood, particularly on the part of the two younger age groups. Overall there was an increase in the mean number of properties related to uses of wood; this increase was 29 per cent, 18 per cent and 12 per cent respectively from the infants, lower juniors and upper juniors. Within these increases, the number of functional properties of wood suggested by infants increased significantly (Wilcoxon Signed Ranks Test, $p<.05$).

Some thoughtful and detailed descriptions of the useful properties of wood were encountered, post-intervention:

> *You can paint it, keep it shiny and shape it whatever shape you want. It*
> *looks better than metal or brick. It's easier than metal to put screws in.*   Y6 B H

Q.	Why do you think wood is a good material for making chairs?
R.	*Because you can paint it all different colours.*
Q.	Right. Any other reason why wood is good?
R.	*Because it will be easier to make it than with steel or metal.*
Q.	Why is that?
R.	*Because you can saw wood and you can't really saw steel so fast.*
Q.	Can steel be sawn?
R.	*Yes.*
Q.	How? Have you seen it?
R.	*I tried it.*
Q.	At home? What were you doing?
R.	*I tried to saw a piece of steel in half and it didn't work.*
Q.	What were you using?
R.	*A metal saw.*
Q.	Do you often try to saw things and make different things?
R.	*No.*
Q.	You just tried that once?
R.	*Yes.*

Y5 B L

Generally speaking, apart from amongst the infants, the gains were fairly slight in the area of relating properties to uses. Both in terms of the science and the technology of materials, it seems likely that with a more focused programme and more direct experiences, greater understanding in this area could reasonably be expected. This is only likely if a range of opportunities and experiences for working with materials is made accessible to children. In the example illustrated by the dialogue above, it seems to be an experience gained at home, and not a particularly successful one, which has informed this child's viewpoint.

As far as some of the other properties of materials are concerned - the economic context and their aesthetic properties, it is unlikely that teachers arranged any very direct input. Here again, there would seem to be scope for development.

5.4   Origins, Manufacture and Changes in Materials

5.4.1   Origins and transformations of materials

Children's ideas about the origins and transformation during manufacturing of two materials was described in section 3.4.1. That discussion will be picked up again at this point, but not before two important cautions are sounded.

The first caution is that the two materials used pre- and post-intervention, although very similar, were not identical. Initially, the stimulus material had been a metal spoon and a piece of coloured cotton cloth; the later discussion centred on metal and cotton again, but this time used the 10 cm length of steel rod and a length of cotton thread. Given the tendency towards context and content-specificity in young children's responses, there is a need to be alert to the influence of the particular form and configuration of the materials presented in each instance and any consequent impact this might have had on the results obtained.

The second caution relates to the mode of elicitation used to obtain children's ideas on each occasion as this also has the potential to influence performance. On the first occasion, the medium through which children were enabled to express their ideas was drawing in picture strip form. On the second occasion, ideas were elicited orally in the course of the individual interviews.

Because of these differences, the data have not been treated as equivalent and have not been presented side by side. However, the data are certainly comparable, not least because of the insights which the differences in the techniques reveal.

Cotton thread

The ideas which children had previously revealed with regard to the small piece of coloured cotton cloth are summarised in Table 3.15 on page 47. Just under a quarter of the sample had traced the cotton back only as far as a product in very similar form - rope, string or thread - in the assumption that the cotton had been unravelled from this 'parent' material. About a third of both the lower and upper juniors had suggested that the cotton fabric originated from sheep's wool. Only about a quarter of upper juniors had indicated an awareness of the origins of the fabric in plant material. The post-intervention data are shown in Table 5.17.

Table 5.17   Origins and transformations of materials: cotton thread

	Post-intervention		
	Infants **n=23**	**Lower Juniors** **n=23**	**Upper Juniors** **n=21**
From sheep	22 (5)	39 (9)	14 (3)
From other animal source	4 (1)	–	–
Cotton 'tree'	–	9 (2)	9 (2)
Cotton plant/'grows'	26 (6)	4 (1)	50 (11)
From rope or string	17 (4)	22 (5)	–
Factory	13 (3)	–	–
Shop/supermarket	4 (1)	4 (1)	–
Other origins	9 (2)	13 (3)	–
'Don't know'	4 (1)	9 (2)	23 (5)

Post-intervention, in response to the length of cotton thread, more of the younger children identified the origins as being with the sheep. This could be interpreted positively as more children beginning to think about origins, so more children making the common erroneous assumption. The number of upper juniors referring to the sheep halved, in the context of the thread.

Fewer children offered the 'unravelling' type of response, relating origins to another, different sized source of thread, located elsewhere; references to shops and factories were at similar low levels (between six and seven per cent) as had been the case earlier. Together, these two types of response, each going no further than the assertion that the product simply exists somewhere else in a slightly different form, ready to be taken 'off the shelf' (and as such, reminiscent of 'cargo cult') accounted for 25 per cent of the sample pre-intervention and 21 per cent (including no upper juniors) post-intervention.

Whereas 10 per cent of the sample revealed awareness of the plant origins of cotton initially, the comparable proportion post-intervention was 32 per cent. It seems unlikely that this particular shift was greatly influenced, if at all, by the different form in which the cotton was presented.

The transformation processes which children referred to are summarised in Table 5.18.

Table 5.18   Transformation processes in production of cotton thread

	Post-intervention		
	Infants n=23	Lower Juniors n=23	Upper Juniors n=22
Planted and/or grown	–	4 (1)	9 (2)
Picked/harvested/cut, etc.	35 (8)	4 (1)	23 (5)
Spinning/weaving	13 (3)	17 (4)	36 (8)
General machine/factory process	17 (4)	13 (3)	5 (1)
Rolled/packaged	13 (3)	4 (1)	9 (2)
Mean number of processes cited per child offering a response of this kind	0.8	0.4	0.8

The processes suggested included planting or growing, from a very small number of juniors only; harvesting, which includes gathering wool from sheep; spinning and weaving (which a minority of children identified as separate processes); more general and unspecified factory processes, and finally, packaging of various kinds. The mean number of processes suggested by children is modest; one process implies two states, so the average response was to suggest just one state prior to the product under consideration.

It might be recognised that the drawings actually suggested a greater experience or more fertile imagination about processes and prior states (see Figs. 3.6-3.8) than is suggested by the data in Table 5.18. This was, indeed, the case. Children generated more states and implied more transformations in their drawings than they made explicit orally in the interviews. However, the drawings made the different *states* explicit rather than the different *processes*. What children were assuming to have happened between the different states which they depicted could only be inferred. On the basis of such inferences, it was apparent that the drawings of the cotton fabric were suggestive of more processes than were the interviews. (The mean number of processes inferred from the cotton fabric drawings were: infants, 0.9; lower juniors, 1.8; upper juniors, 2.5.)

Metal rod

The number of children who mentioned ores, rocks, mines or unformed metal under the ground was minimal in the pre-intervention drawings of the origins of the metal spoon (see Table 3.16). Two-thirds of the sample suggested that the spoon had been fabricated from an existing source of metal, either recycled or existing in storage ('on the shelf'). The post-intervention data emerging from the individual interviews centred on the metal rod are summarised in Table 5.19.

Table 5.19   Origins and transformations of materials: metal rod

	Post-intervention		
	Infants **n=23**	**Lower Juniors** **n=23**	**Upper Juniors** **n=22**
From ore or rocks	–	4 (1)	18 (4)
From mine/underground	4 (1)	–	–
Unformed metal in or on ground	4 (1)	–	18 (4)
Originally metal in liquid form	–	26 (6)	23 (6)
From metal in another shape with another property	26 (6)	26 (6)	9 (2)
Recycled from scrap/ from the tip	13 (3)	13 (3)	14 (3)
From a factory	17 (4)	–	–
From a shop	4 (1)	–	–
Always existed in present form	4 (1)	–	–
From another method	9 (2)	–	–
Other sources	9 (2)	13 (3)	–
Don't know/no response	9 (2)	17 (4)	14 (3)

The proportion of children indicating that the source of metal was the Earth's crust - more specifically, in or on the ground in an unformed state - was 16 per cent post-intervention, compared with 12 per cent initially.

Q.   What did it look like before this?
R.   *It was like a liquid before, heated in a big pot.*
Q.   And before that?
R.   *It was a stone on the hills.*   Y5 G L

There was no evidence of any explicit understanding about metal ores. Those children who were aware that the metal existed in an earlier state in the ground tended to think of it as much the same material but in a different location.

> *It was like a lump, buried in the ground, all dirty.*   Y6 G H

As children are likely to be depending on secondary sources for this sort of information, the references to mining metal which they are most likely to have encountered will have been gold or silver mining, in which miners strike veins or nuggets of pure gold, or 'pan' for the pure form of the metal in river beds. A mine in their locality might possibly lead to children being familiar with more complex methods of extraction, but this is unlikely.

There was an increase in the number of children suggesting that the original condition of metal would have been a liquid state (three upper juniors initially, five post-intervention plus another six lower juniors).

> *It came from a silver puddle. It was frozen and made into the rod.*   Y3 G L

> *It was liquid, made into the rod. It was put in the oven to get it hard.*   Y3 B H

> *It was a sort of liquid, like mercury.*   Y3 B L

> *It was a kind of liquid. It was left to cool and it hardened.*   Y5 G L

It would be difficult, even for a chemist or mineralogist, to describe an 'original' state of the metal rod, whatever that might be, and it is not possible to be absolutely clear whether children were referring to a transformation process or an 'original', i.e stable, condition in which metal is to be found prior to any manufacturing process. Whereas the fourth response quoted above is consistent with the idea of a molten state during a manufacturing process, it seems less likely that the other three comments rest on similar assumptions. The first response above carries the notion that metal is subjected to a freezing process to harden it; the second suggests something like a baking; the third refers to a metal which exists in a liquid state at normal ambient temperature; the fourth probably refers to heating as a means of changing the state of metal from solid to liquid as part of a manufacturing process. All of these attempts to make sense of the change of state of metals generalise from other knowledge or experience. Although the use of metal in the production of everyday artefacts is so widespread - cars being just one example - the manufacturing processes in their production are not accessible to children. The blacksmith's forge is history; heritage centres might be a source of information to some children: it is still possible to see iron being cast in such centres. For most, the transformation processes in one of the most basic and important of twentieth century manufacturing materials are hidden from direct experience, available only through secondary sources.

In the light of the responses and discussion presented above, it should probably not be surprising to find the majority of children locating prior states of the metal rod in the form of existing metal having another shape or property, or alternatively, as having been re-cycled from scrap. Twenty-one per cent of responses were of this 'metal from another source' variety (67 per cent in relation to the pre-intervention spoon); re-cycling responses were at an overall level of 13 per cent, compared with nine per cent in relation to the spoon.

It was from the tip where they crash cars and break them into pieces. Y3 B M

The transformation processes which children suggested are shown in Table 5.20.

Table 5.20   **Transformation processes in production of metal rod**

	Post-intervention		
	Infants **n=23**	**Lower Juniors** **n=23**	**Upper Juniors** **n=22**
Reference to heating	17 (4)	4 (1)	–
Reference to heating to soft molten or liquid state	9 (2)	39 (9)	64 (14)
Reference to moulding	4 (1)	–	14 (3)
Shaping/rolling/ crushing process other than moulding	30 (7)	17 (4)	18 (4)
More general reference to machine/factory	17 (4)	17 (4)	5 (1)
Mean number of processes cited per child offering a response of this kind	0.8	0.8	1.0

A distinction is made between those comments which indicated a role for heating in the production of the metal rod and those which more clearly suggested that the effect of the heating was to alter the plasticity or state of the metal. Almost two-thirds of the upper juniors mentioned heating. The younger children showed less awareness of this process.

As with the cotton thread, the mean number of processes cited per child was fairly modest, and fewer than could be inferred when the responses were elicited in the form of drawings.

5.4.2   Ideas about the possibilities of transforming metal

More focused questioning was used in this area. The metal rod, a piece of metal wool, a short length of wire, metal powder (fine filings), and a piece of metal foil were presented. Children were then asked whether it was possible to change the metal rod into each of the other forms, and if so, how. The results for the metal wool were reported in section 3.4.2, as questioning in relation to that item was not repeated, post-intervention.

As reported for the metal wool, responses were categorised according to whether they made mention of heating, a mechanical process or both of these. In those cases in which children denied the possibility of the transformation taking place, frequencies of reference to this being because of different attributes, being different materials or different metals were recorded. Table 5.21 summarises the possibilities put forward by children for the transformation of the metal rod to wire.

Table 5.21   Ideas about the possibility of transforming metal: rod to wire

	Pre-intervention			Post-intervention		
	Inf n=23	LJ n=23	UJ n=22	Inf n=23	LJ n=23	UJ n=22
Transformation deemed possible described in terms of:						
Use of heat	17 (4)	4 (1)	18 (4)	–	13 (3)	23 (5)
Use of machine/ mechanical process	22 (5)	30 (7)	23 (5)	52 (12)	9 (2)	18 (4)
Use of heat *and* mechanical process	–	13 (3)	36 (8)	4 (1)	9 (2)	27 (6)
No specified method	4 (1)	–	5 (1)	9 (2)	13 (3)	14 (3)
Total suggesting rod to wire transformation was possible:	44 (10)	48 (11)	82 (18)	65 (15)	44 (10)	82 (18)
Transformation deemed *not* possible, because:						
The two objects had different attributes	35 (8)	44 (10)	–	13 (3)	30 (7)	–
The rod and wire were different materials	–	–	5 (1)	–	–	–
Only 'magic' would permit transformation	9 (2)	–	–	4 (1)	–	–
Other reasons	–	–	–	4 (1)	9 (2)	–
Total suggesting transformation *not* possible	44 (10)	44 (10)	5 (1)	22 (5)	39 (9)	–
No response/don't know	12 (3)	9 (2)	13 (3)	13 (3)	17 (4)	18 (4)

More children deemed the transformation possible than not possible and this affirmation was slightly increased amongst the infants, post-intervention. There was also a slightly increased reference to the mediating role of both heat and mechanical processes, but not in the frequency of response in which both of these processes were suggested.

The fact that the two metal products had superficially different attributes appeared to be less of a barrier to accepting the possibility of converting one to the other, in the follow-up interviews.

It was interesting to note that the possibility of changing one form of metal to the other was so remote to one or two of the younger children that they explicitly invoked the idea of magic.

Q.   Would it be possible to change the metal rod into wire?
R.   *No, you've no magic*   Y2 G M

Table 5.22 presents parallel data for the possibilities of transforming the metal rod into powder.

Table 5.22   Ideas about the possibilities of transforming metal: rod to powder

	Pre-intervention			Post-intervention		
	Inf n=23	LJ n=23	UJ n=22	Inf n=23	LJ n=23	UJ n=22
Transformation deemed possible, described in terms of:						
Use of heat	9 (2)	9 (2)	–	–	17 (4)	–
Use of machine/ mechanical process	26 (6)	48 (11)	73 (16)	70 (16)	26 (6)	86 (19)
Total suggesting rod to powder transformation was possible:	65 (15)	61 (14)	82 (18)	70 (16)	52 (12)	91 (20)
Transformation deemed *not* possible, because:						
The two objects had different attributes	26 (6)	17 (4)	5 (1)	9 (2)	9 (2)	5 (1)
The rod and powder were different materials	9 (2)	4 (1)	5 (1)	4 (1)	–	–
Other reasons	–	9 (2)	–	–	9 (2)	–
Total suggesting transformation *not* possible	35 (8)	30 (7)	9 (2)	13 (3)	17 (4)	4 (1)
No response/don't know	–	9 (2)	9 (2)	17 (4)	31 (7)	5 (1)

Of the three transformations discussed in this section, rod to powder was the one most widely deemed to be possible. The most commonly cited process was a mechanical one involving chopping, grinding or sawing.

By sawing, and all the little flakes are dropping. Y6 B M

The powder is crushed metal. Y5 G H

Cut it up with a sharp machine. Y6 B M

An alternative idea was that the application of heat, or the process of burning, would be sufficient.

Burn it and burn it and burn it. Y3 B H

Melt it and get little bits off and let them set in a hot place because sometimes it sets better in a hot place and sometimes in a cold place. Y3 G H

The possibility that the metal rod might be transformed into the foil was difficult for most children to accept (18 per cent of the sample thought it might be possible, compared with 71 per cent for rod to powder and 63 per cent for rod to wire, post-intervention).

Table 5.23   Ideas about the possibility of transforming metal: rod to foil

	Pre-intervention			Post-intervention		
	Inf n=23	LJ n=23	UJ n=22	Inf n=22	LJ n=23	UJ n=22
Transformation deemed possible, described in terms of:						
Use of heat	4 (1)	4 (1)	9 (2)	–	–	5 (1)
Use of machine/ mechanical process	13 (3)	4 (1)	–	18 (4)	22 (5)	9 (2)
Use of heat *and* mechanical process	–	4 (1)	18 (4)	–	–	–
No specified method	–	–	5 (1)	–	–	–
Total suggesting rod to foil transformation was possible	17 (4)	13 (3)	32 (7)	18 (4)	22 (5)	14 (3)
Transformation deemed *not* possible, because:						
The two objects had different attributes	17 (4)	13 (3)	27 (6)	41 (9)	26 (6)	9 (2)
The rod and foil were different metals	–	–	5 (1)	–	–	5 (1)
The rod and foil were different materials	13 (3)	48 (11)	14 (3)	14 (3)	26 (6)	31 (7)
Other reasons	–	–	–	–	4 (1)	–
Total suggesting transformation *not* possible	31 (7)	61 (14)	45 (10)	55 (12)	56 (13)	45 (10)
No response/don't know	52 (12)	26 (6)	23 (5)	27 (6)	22 (5)	40 (9)

The kind of metal foil used was typical aluminium kitchen foil which would be familiar to most children either in the form used or as the 'silver paper' used to wrap sweets or chocolate. Only one child was clear that the foil was not metal:

> *Foil isn't proper metal. It's not strong.*   Y3 B L

Another was less certain, and seemed to have a change of mind:

> *Foil is not metal. If it was metal, you could shave it down.*　　　Y6 B H

Only one child recognised that the steel and aluminium foil are different metals.

Twenty-four per cent of children interviewed asserted that the metal rod and the foil were different materials. This suggestion increased with age.

> *No. It's not made from metal in the first place.*　　　Y6 G M

> *It's a different thing altogether.*　　　Y5 G H

> *The foil could have been a piece of paper already, because it came from trees.*　　　Y6 G M

> *Metal foil is made from paper. You cut it and dip it into silver, to make the colour.*　　　Y6 G M

It perhaps needs to be emphasised that the foil used was not that which is often backed with paper. It is clear that a number of children just did not perceive foil as being a metal. The influence of the commonly used term 'silver paper' may be to blame.

An approximately equal number of children, 25 per cent of the sample, focused on the different attributes of the two items as being the explanation for the impossibility of the transformation; this type of reasoning decreased with age.

> *No, it would have to be a flatter silver. If it was just brown, people would think it isn't foil because foil's always silver.*　　　Y2 B M

The 'silver' is treated by this young child as a pigment rather than a reflective property, reminiscent of the way in which the transparency of water is often described as 'white' by young children. Although it is a superficial property which has seized this child's attention, the experience and understanding implied in avoiding such an apparently reasonable conclusion is formidable, though not impossible. (Activities with metal polish suggest themselves.) Some other, more immediate perceptual properties also enticed children into asserting that the transformation from rod to foil would not be possible:

> *Foil is a lot lighter.*　　　Y3 B M

> *The foil is soft and the rod is hard.*　　　Y2 B L

These sorts of properties are probably more amenable to some sort of empirical investigation.

138

6. SUMMARY

6.0   Introduction

This report presents a series of related investigations into children's understanding of materials. The work was conducted collaboratively with twelve teachers in six schools; within the classes of the participating teachers, all children contributed to the research. Sixty-eight children were randomly selected to form a stratified sub-sample. These children were interviewed at the end of an exploration phase during which they were given the opportunity to familiarise themselves with a range of materials. Their ideas were elicited and summarised and a similar interview procedure was repeated about eleven weeks later following a period during which they were exposed to a range of classroom-based intervention activities.

The conceptual issues around which the study was structured were the following:

Descriptions and properties of materials

Solids, liquids and gases

Uses of materials

Origins, manufacture and changes in materials.

Within each of these issues, a number of themes were identified. The following paragraphs present a brief summary of the main points of interest emerging within each of the four issues.

6.1   Descriptions and Properties of Materials

6.1.1   Classification of materials

Children were asked to put a range of materials into sets, according to what they were made of, i.e. their composition. In examining responses, it was found that the kinds of set definitions which children used could usefully be divided into five types:

Compositional
Functional
Locational
Perceptual
Manufactured

Roughly half of the sets invented by children actually referred to the composition of materials, suggesting that the task of classifying objects according to what they are made of was found to be challenging. Almost half the set labels produced by infants related to perceptual properties - 'hard', 'soft', 'shiny', etc. This type of perceptually dominated response was far less common amongst the juniors.

After compositional properties, the next most common type of set used by the juniors referred to the function of objects - 'for building with', etc.

The locational type of set label was found amongst the juniors, but only infrequently. It referred to a common location such as 'found on the beach'.

The set label 'manufactured' was recorded because it was regarded as potentially important in the classification of materials. It was virtually absent from children's responses.

The intervention consisted of observation and other investigational activities, but not of a systematically organised nature. Children's set labels showed stability before and after this intervention; the most frequently used set labels were the same before and after the intervention. Despite being represented by a heterogeneous group of objects, metal objects were classified relatively successfully.

6.1.2   Judgements about the properties of materials

The active exploration of various properties of materials was central to the classroom activities and the ways in which children made judgements about the harder and stronger of two materials have been summarised. It was noted that some children were satisfied with observational evidence, while others implied the use of a test in coming to a decision. The proportion of children referring to such tests increased after intervention, though it became clear that the particular materials under consideration had an important bearing on the kinds of evidence and the kinds of tests which children were likely to suggest. It seemed that infants in particular demonstrated an increased likelihood of perceiving the relevance of some sort of test to their decision making when comparing properties of materials.

6.2   Solids, Liquids and Gases

Children's ideas about how smelling occurs were recorded, without any direct attempts to shift their ideas in this area. Some children, more often the younger ones, tended to focus on smelling as an active process, mentioning the role of the nose only, apparently assuming that to mention the nose was sufficient to explain the process. (There are parallels with notions of 'active seeing' and 'active hearing', reported in SPACE research reports on Light and Sound, respectively.) Other children referred only to the material or its properties, as though these had an intrinsic property of 'smell'; between a quarter and a third of children at all ages showed at least the beginnings of an understanding of the interaction between the object smelt and the organ of smell, though there was no evidence of intuitive particle theories.

Children's ideas about the air were explored using an 'empty' container, with a counter-suggestion that some opinions held that the container was not, in fact, empty. Although almost universally, children's first reaction had been to treat the container as empty, following the counter-suggestion, most of the juniors suggested that it would contain air. Less than a quarter of the infants made a positive response to the counter-suggestion by referring to air.

The frequencies with which various characteristics of air were suggested by children - invisibility, being life-sustaining, being associated with felt movement, and less frequently, odourlessness and relatedness to temperature sensation - are reported.

Older children showed evidence of an increasing awareness that air is to be found all around.

When asked to produce drawings of solids, liquids and gases, the latter were in the minority; the number of gases represented in drawings post-intervention increased, but the age-related trend remained in evidence. Many children's examples of gases were triggered by the association with the word 'gas', as in 'gas-cooker'. The range of defining attributes of gases revealed through discussions with children are reported.

The solids which children chose to draw tended to be strong and hard; after the intervention, children were much more likely to include soft solids and powders in their illustrations.

The liquids which children included in their drawings were frequently suggested by associations such as 'washing-up liquid'. More viscous liquids which had tended to be classified as 'neither solid nor liquid' pre-intervention, were increasingly likely to be accepted as liquids after the period of classroom activity. This difficulty with viscous liquids is probably explained by the fact that 'runniness' was treated by the majority of children as being a critical attribute of any liquid.

Children's classifications of five materials (metal rod, cotton wool, vinegar, treacle, talcum powder) as solid or liquid were seen to have become significantly more accurate in scientific terms after the intervention. The shifts in the responses of individuals are reported.

Children's ideas about the identity of a colourless liquid within a sealed container were investigated. Their initial response, almost unanimously, was that the liquid was water. When asked what else it might be, most of the juniors readily suggested other possibilities, though three-quarters of the infants did not. The most common assumption was that the colourless liquid could be some other form of beverage. Children showed minimal awareness of safety issues related to smelling or tasting unknown substances.

6.3   Uses of Materials

Children were asked to relate the properties of four materials - wood, rubber, wool and metal - to their use in fabricating particular objects. Responses were classified according to four criteria:

Functional properties
Manufacturing properties
Aesthetic properties
Economic properties

142

These particular properties had not been targeted by teachers, though there was intervention of a more general nature. Functional properties were most frequently mentioned, followed by manufacturing considerations. Aesthetic properties were mentioned infrequently and economic considerations hardly at all. Apart from amongst the infants, gains were not great. Children expressed some interesting ideas in this area, which links closely to design and technology. There was also the impression that, given more direct experiences, further gains would be possible.

6.4 Origins, Manufacture and Changes in Materials

6.4.1 Origins and transformations of materials

Drawings and interview discussions were the two modes used for collecting data in this area and the differences associated with each method are discussed. Differences in performance associated with the consideration of different products made of essentially the same materials are also discussed.

A number of children confused the origins of cotton and wool; others suggested that cotton was obtained from a source such as rope or string, through a process of unravelling. Generally speaking, children did not show a great deal of awareness about the world of manufactured objects.

Most children thought that metal objects were made from metal which was available in another form or in another place; for a minority, this source was recycling. Only 16 per cent of children indicated some sort of awareness that the origins of metal were in or under the ground or in some way associated with the Earth's crust, post-intervention. First-hand experiences in this area were, understandably, not available.

6.4.2 Ideas about the possibilities of transforming metal

A length of metal rod, a piece of metal wool, a short length of wire, metal powder and a piece of metal foil were presented to children and they were asked about the possibility of transforming the metal rod to each of the other forms. The rod to metal wool and rod to foil transformations were those which children were least confident about treating as possibilities. Many children did not regard metal wool as being made of the same material as the rod. The rod to wire and rod to powder transformations were deemed possible by the majority of children interviewed.

Some indications of the processes which might be involved in transforming metal from one form to another were also recorded, particularly references to the role of heat and the part played by mechanical processes. Children lack experience in this area and, consequently, many confusions about how these transformations of metal occur were encountered.

6.4.3   *Predicted changes on heating*

Pre-intervention only (since no classroom-based activities in this area had been planned), children's assumptions about expected changes when the metal rod, cotton wool and vinegar were each separately heated were established.

The majority of juniors and just under half the infants showed awareness of a change in at least the plasticity of metal when subjected to heating. About half the interview sample predicted that there would be a net mass decrease in the metal rod as the result of heating.

A quarter of the infants suggested that the cotton wool would become hotter but there would be no other change if it were to be heated to a very high temperature.

Forty per cent of the interview sample suggested that the vinegar would stay the same when subjected to a very high temperature. A minority of children at all ages mentioned the possibility of the vinegar boiling, bubbling or foaming; this kind of prediction increased with age to include about one-third of the upper juniors.

6.5   *Concluding Comments*

Reviewing the work on children's understanding of materials as a whole, a few more general comments are pertinent, to help to set this research within the context of other SPACE Project research into children's understanding of key concepts in science.

The first point to make is that there is not such a clear boundary between the everyday domain of thinking and expressing ideas about some aspects of materials and the more strictly scientific domain, for the age group under consideration. A topic such as 'materials' might be thought of as more widely embedded in everyday meanings and contexts than might be the case for some other topic areas - magnetism and electricity, for example. The majority of materials to which children were exposed were everyday ones; they could safely be observed and manipulated and, in many cases, had been experienced in a variety of forms and contexts. There is no problem about making a large range of materials accessible to children. On the other hand, some states or forms which are particularly interesting because they exemplify some core concepts or key ideas are extremely inaccessible - changes of state of a range of materials, for example. The knowledge of materials assembled from everyday sources consequently provides a very uneven basis for transition and elaboration from the vernacular to the science domain.

On reflection, children's lack of knowledge about many materials is understandable. The origins of many materials are obscure. Many manufacturing processes are hidden from view; they may take place in factory environments which are inherently difficult to visit. Opportunities to help children to develop their understanding of the origins of materials and the changes that occur in them during manufacturing processes, through direct experience, may therefore be limited. When knowledge is acquired, it is often second-hand, lacking the immediacy of direct experience which teachers strive to make available to children. In the

144

absence of direct experience, naturally enough, young children simply accept the world as existing in the form in which they encounter it. Manufactured items are found in shops or come from factories. There appears to be insufficient challenge to children to reflect upon the fact that objects have not always been in the form in which they come into the home, that they can change in a variety of ways through a multiplicity of processes.

Another important point is that whilst many material objects are familiar, they are not necessarily thought of by children in terms of what they are made of; the quality of familiarity is more likely to reside in their surface features (colour, shape, size) or in their uses (to eat, to play with, etc.) than in their composition. Although the 'material' of which an object is composed might be thought to be a very concrete property (for example, in the sense of being tangible), it is not necessarily the most discernible or most readily perceived feature. The material of which an object is made may often be obscured by other considerations, particularly function. To complicate matters further, there is a well-developed technology employed to disguise or embellish materials by treating their surfaces. The nature of naturally occurring materials or manufactured objects can be brought to awareness, given appropriate opportunities and experiences; such experiences extend everyday experience into the domain of technology. The interviews tended to suggest that many children lacked these more exploratory investigations of the properties of materials, other than at home. Given the opportunity to explore systematically the properties of a wide range of materials, gains might be anticipated in this area; an opportunity also exists for the Design and Technology curriculum to complement the science curriculum in this particular respect.

Even within the everyday domain, there were obvious problems with the basic vocabulary of properties and states of materials. There was no attempt to resolve such problems in the simple-minded manner of giving children the right words. What teachers did manage to do with particular success, especially with the younger age group, was to provide examples of materials, experiences of transformations, opportunities for discussion and so on, which enabled children to extend or otherwise modify the boundaries of the definitions which they were operating.

List of Appendices

Appendix I   SPACE School Personnel (for research carried out in 1990)

Appendix II   SPACE School Personnel (for research carried out into
                    children's classification of materials in 1987)

Appendix III   Exploration Experiences

Appendix IV   Pre-Intervention Interview Schedule

Appendix V   Interviewing Techniques: General Guidelines

Appendix VI   Intervention Guidelines

Appendix VII   Post-Intervention Interview Schedule

Appendix VIII   Bibliography

APPENDIX I

SPACE SCHOOL PERSONNEL

(for research carried out in 1990)

Lancashire L.E.A.

County Adviser for Science: Mr P. Garner

Advisory Teacher: Mrs R. Morton

School	Head Teacher	Teachers
Brindle Gregson Lane Primary	Mr P. Maddison	Mrs O. Beal
Frenchwood Primary	Miss E. Cowell	Mrs L. Bibby Mrs C. Pickering
Longton Junior	Mr M. P. Dickinson	Mrs S. Hamer Miss R. Hamm Mrs L. Whitby
Scarisbrick Primary	Mrs S. Harrison	Mrs S. Harrison Mrs A. Stocks
St Mary and St Benedict R.C. Primary	Mr M. Sugden	Mrs F. Flowers Mr A. Hoyle
Walton-le-Dale Primary	Mr A. D. Roberts	Mrs J. Hewitt Mrs M. Truscott

APPENDIX II

SPACE SCHOOL PERSONNEL

(for research carried out into children's
classification of materials in 1987)

Lancashire L.E.A.

County Adviser for Science: Mr P. Garner

Advisory Teacher: Mrs J. Knaggs

School	Head Teacher	Teachers
Banks St Stephen C.E. Primary	Mr N. McMechan	Miss L. Iddon
Hillside Primary	Mr J. R. Worthington	Mrs R. Hobbs Mrs A. Joyce
Holland Moor Primary	Mr M. Beale	Mrs P. Lock Mrs L. Pinnell
Ormskirk C.E. Primary	Mr J. Rigby	Mrs J. Myers Mr D. Sharkey
Town Green Primary	Mrs C. M. I. Thornton	Mrs J. Lofthouse Mr M. Taylor
West End Primary	Mr P. Guy	Mrs S. Harrison Mrs R. Wallington

APPENDIX III

EXPLORATION EXPERIENCES

Display

Collect and display everyday materials. The list below indicates a range of different kinds of solids, different kinds of liquids and gases. Please do not organise the display in any way which reflects possible ways of classifying the materials. At this stage we have to avoid giving the children words (such as solid, liquid, or gas) in order to find the words that they naturally use. Keep the materials in their usual containers but have other containers available so that children can experience those things that are hidden by their packets. On a wall behind this display, arrange cards on each of which a question is written:

> *What do you know about each material?*
> *What do you notice about each material?*
> *What could you do to any of the materials to find out more?*

These questions are intended to be prompts for preliminary thoughts and activity rather than any detailed investigation (more suitable for the intervention phase of research).

Here is the list of materials; make sure that materials have labels if it is not obvious what they are.

tomato sauce
honey
paint (thick, e.g. gloss)
cooking oil
lubricating oil (e.g. 3 in 1)
orange squash
washing-up liquid
water
salt
coffee
flour
washing powder
piece of coloured cotton fabric
sponge
grease
plasticine
kitchen foil
balloon (deflated)

plastic bag
piece of clear plastic
polish (i.e hard, wax-type variety)
piece of rigid plastic (e.g. tray)
plank or other piece of wood (that has been manufactured from the original wood)
brick
piece of rock
spoon (an old one you don't mind being damaged)
bottle of coke (or other 'fizzy' drink)
foam (any kind which has bubbles in it)
'aero' chocolate bar
balloon (blown up)

150

Safety

If you feel that any of these materials are potentially unsafe, then please control access to them, perhaps keeping them with you and only letting children see them with you there. You might be able to provide safety goggles and gloves when children are treating the materials and finding out about them. Warn children about the dangers of smelling, tasting or even touching unknown materials.

ACTIVITIES and PRODUCTS

Please ensure that names and dates are on each piece of work.

Activity 1 Comments on Display Materials

Either let children keep their own individual records about any of the things in the display. They should use the following format on a large sheet of paper, with each child having their own sheet of paper.

Materials	What I can say about it	Date

Or if you prefer, ask children one at a time to choose an item from the display and come and tell you what they notice about it. You could then write down the comments for them, perhaps using magazine pictures instead of children's pictures/words in the first column on a chart which you can then display.

Product

For each child, a record of their comments either as their individual comments written on paper or comments written on behalf of the child by the teacher.

Activity 2 Finding Out About Materials

Have a box with a set of 'implements' which children can safely use on materials. You might have to select according to the age of your children but it could include:

> water and suitable containers plus spoons
> a file
> something to squeeze/bend materials like a pair of pliers
> a torch
> a magnet
> sandpaper
> a sieve
> a pestle and mortar (or two stones to rub things between)
> pair of scissors
> hammer and nail
> magnifying glasses
> moulds (e.g. biscuit moulds, sand moulds)

Ask individual children to select one of these implements (or they might even use their hands as an implement for pressing, squashing etc.) and a range of materials. For safety reasons, you may want to supervise the materials they collect. Invite them to find out what they can about these materials. Ask them to make a record of what they did and found out. (Note: for safety reasons, we thought it best to exclude heating materials at this stage; it would fit better as an intervention when the range of materials for heating can be more carefully controlled.)

Product

For each child, a piece of paper showing what they did and found out (written by themselves or recorded by the teacher as necessary).

Activity 3 Where Does It Come From?

Choose a relatively easy example like the plank of wood and show the children drawings of what it is like, what it was like before that (a bigger piece of wood) and before that (a log) and before that (a tree). Ask them to make similar drawings for:

1   flour
2   piece of coloured cotton fabric
3   spoon

(Remember that all drawings tell us more about children's thinking if they are annotated, i.e. the child (or the teacher on behalf of the child) writes comments explaining the drawing or parts of it. It is not neatness or artistic merit that we are after here but clues as to the child's ideas. Drawings without words are often difficult to interpret).

Product

For each child, an annotated drawing for the three materials. It might be in the form:

What it is like	What it was like before that	And before that	And before that
Flour			
Coloured cotton fabric			
Spoon			

Putting four spaces next to each one does not imply four drawings are required in each instance. The child has to be encouraged to go back as far as seems reasonable to him/her.

154

Activity 4 Comparing Materials

Ask the children to take container A and say how they think the contents are alike and how they are different from each of the contents of the containers B, C, D and E. They should not open the containers but they should be able to look carefully and tip the contents if they wish.

Recording

Some children may be able to form their own tables to show their observations.

Pair of containers	Alike	Different
A and B A and C A and D A and E		

Otherwise you could talk to children individually (restrict yourself to the sub-sample, if necessary) and write what they say for them.

Product

For each child (or the sub-sample, at least) a table showing their comparisons of the materials in the containers.

Activity 5 Ideas About Solid, Liquid, Gas
(must come after Activity 4)

Let pupils draw pictures showing solids, liquids, and gases. Let them come up with their own ideas in response to these three terms. Their drawings might be done on a page divided into three sections. Where the words appears to mean nothing to the child, they might leave that section blank. Older children may want to list further items as well as making a drawing but let them start with drawings first. Drawings should be labelled by the child or the teacher.

Product

For each child, a drawing showing his/her ideas of things that are solids, things that are liquids, and things that are gases.

APPENDIX IV

PRE-INTERVENTION INTERVIEW SCHEDULE

Types and Uses of Materials

A   **Meaning of Hard and Strong**

If you choose any two of the materials (invite child to specify) how would you decide which was:

1.   harder

2.   stronger

(Try to ascertain child's concept of these two terms.)

For younger children you may need to drop back to the question 'Which is harder?' using concrete examples.

B   **Relating Properties to Uses**

3   Why do you think wood is a good material for making furniture?

4   Why do you think rubber is a good material for making tyres?

5   Why do you think wool is a good material for making clothes?

6   Why do you think metal is a good material for making coins?

(Probe to see if the child can offer more than one reason in each case.)

156

C   **Concept of Solid and Liquid**

Use the materials A-E, originally used as part of exploration experiences

Identify: A as steel,
B as cotton wool
C as treacle
D as talcum powder
E as vinegar

Ask: 7   Which do you think are solids?

8   How did you decide?

9   Which do you think are liquids?

10   How did you decide?

11   For any left over, ask: 'So what is this?'

D   **Concept of Gas**

Remove lid from E (vinegar) container. Let child smell the vinegar. Ask:

12   'How are you able to smell the vinegar?' and/or 'What happens to let you
smell the vinegar?' to probe whether they recognise movement of vapour from
vinegar to their noses.

13   Show 'empty' container. Say that someone says it is not empty - why do you
think they say that?
What do you think?

14   Probe knowledge of properties of air, e.g. What can you tell me about air?
Do you think it weighs anything? What do you think air is? (Do you think
air is a gas?)

15   Probe knowledge of extent of air. Where can you find air?

E   **Changes on Heating**

Imagine that we put each of these, the steel rod, cotton wool, vinegar into a small open container that you can heat to a very high temperature - as high as you like:

16   What do you think would happen to this one (A)?

17   Probe mass conservation (before and after heating), e.g. Suppose you weighed it now and then after you heated it ...

18   What do you think would happen to this one (B)?

19   What do you think would happen to this one (E)?

20   Probe location, if suggested, of vapour.

F   **Colourless Liquid**

Show container with colourless liquid.

21   What could this be?

22   What else could it be?

23   Probe how you could find out what it is, e.g. Someone said it wasn't water, how could you find out?

G   **Transforming Metal - Processes Involved**

Use: steel rod (A)
wire
foil
wool
filings

If you started with the metal rod, could it be made the same as this? How? (Probe also why not, if say no, to see acceptance of the five materials as metals.)

24   Metal wire

25   Metal wool

26   Metal foil

27   Metal powder

APPENDIX V

INTERVIEWING TECHNIQUES: GENERAL GUIDELINES

1. In order to explore children's own concepts it is important that their ideas should not have been influenced by teacher input. It is also important that the interviewer has no preconception about the content of the child's responses.

2. The child must be aware of the purpose of the interview, which is to establish the child's ideas about a particular topic. It must also be clear to the child that the interview is being recorded, either by transcription or on audio tape.

3. The interview questions are there as a guideline. One question may be enough stimulus for a whole interview. It may be more informative to follow the child's line of thinking than to try and go through the questions from beginning to end. However, it should be ensured that the concepts the questions focus on are adequately covered.

4. Questions should be non-directive. They should be phrased so that the child can give any answer they feel is appropriate:

    Tell me what you know about clouds.

5. The child should be allowed to talk freely, and the interviewer's attention given to picking up any critical words or phrases. These remarks can then form the basis of further questions which might focus more closely on the child's ideas.

    We have to wash the beans twice a day because they need water or they'll shrivel up.

    Can you tell me why you think the beans need water?

6. The child should be encouraged to expand any answer they give, even if they appear to have contradicted previous statement. A look which registers the interviewer's surprise may inhibit the child's elaboration.

7. The questions should require no factual knowledge on the part of the child.

    What is the name of this plant?

    is not saying anything about the child's conceptual knowledge of plant growth.

    What can you tell me about this plant?

    might elicit the same information and also allow for other ideas to be put forward.

8.   Questions with a yes/no answer should be avoided because they do not allow a free response.

        Do you think the glass is cold?

could be better phrased as:

        Have you any ideas why this happens to the glass?

9.   It is a good idea to repeat what the child has said to make sure:

    a.   that they have meant what they said

    b.   that their intended meaning is what has been understood.

Recapping may also prompt the child to elaborate their answer.

10.   It is possible that the interviewer and the child may use the same word to mean different things. It is necessary to ensure that the child's everyday meaning is understood.

11.   The length of the interview will obviously vary both with the subject matter and the child. It is important, though, that the discussion is not extended beyond its natural time limit in the hope of obtaining new ideas.

12.   During the course of an interview, children may change their minds. When this happens, *all* the ideas expressed should be recorded, in sequence. The temptation to summarise the discussion in terms of the final idea expressed will give a false impression of commitment or certainty, and should be avoided.

13.   All the above points stress the non-directive nature of the data collection. At the same time, interviewers must keep in mind the focus of interest of the interviews. This means:

    a.   all the points of interest to the topic are covered

    b.   not too much time should be spent on drifting into interesting but non-focal areas.

APPENDIX VI

INTERVENTION GUIDELINES

Types and Uses of Materials: Intervention Phase

The Exploration Phase activities and interviews suggested several possible areas for intervention. In order to make intervention manageable, the following activities focus on:

1.   Thinking about why particular materials have been chosen for the job they do.

2.   Vocabulary work, based on real objects helping children to

   i   describe materials, and
   ii   distinguish solids, liquids, gases.

3.   Where materials have come from and the various transformations that they have undergone.

4.   i   Investigating the behaviour of different materials
        ii   seeing how various materials can be changed.

162

Activity 1 Materials in Use around the School

Let children in groups try to identify different materials in use in and around the school. These might be both constructional materials - brick for building, glass for windows, etc., and others - plastic for containers, chalk for writing, etc.

Each group might form a table in words or pictures as below:

Materials	Use

Then let them discuss in their groups (or as a whole class) why they think each material was chosen for its particular use. Use questions of the type 'What makes chalk a good material for writing?' If the children are still working in groups, they might come to an agreement through discussion and add their agreed reasons to their tables.

Product

For each group, a table of materials, their use and reasons why the material fits its use.

Activity 2 Vocabulary Work

Encourage children's language development in relation to actual materials and experiences of those materials. The following were suggested in the meeting and you may like to try these or to substitute others. Please do some work on describing materials and some on the idea of solids, liquids and gases.

a. Using sense of touch

Use materials from the original display as appropriate. Blindfold children and let them feel materials and describe them.

b. Guessing game

Let one child describe a material to others in a group (or class), perhaps one item of description at a time so that the others can work out what it is. Encourage children to focus on the material itself rather than its package or use or other information about it.

c. Solids, liquids, gases

Choose from these terms what you think would be appropriate for your age group. For example, gases might be suitable only for upper juniors. Again you will have to judge when children are ready to distinguish between using solid to describe something as rigid and firm (e.g. a solid table) or as being the same all the way through. (i.e. non-hollow as in solid rock) or a solid substance (which includes soft things and powders).

Let children find examples of, for example, liquids. These might be displayed together as the real material, where possible, and pictures, where not.

Encourage children to generalise, e.g. not to restrict their idea of liquid to colourless or watery ones.

You might then ask children to talk about the examples in their display, prompting them to say what they all have in common (in order to develop good describing words for liquids rather than a definition).

For gases, experiences might be given of blowing bubbles, making ginger beer, using spray cream aerosols, perfume, 'atomisers', blowing up balloons, helium balloons, camping gas cookers, deflating bicycle tyres, shaking lemonade containers!

Product

Some account in the intervention diary of vocabulary work undertaken with any observations you have on it.

164

Activity 3 Where Does It Come From?
(Please use materials kit if you wish)

Let children trace back materials to their origins. They can find out the various stages in the manufacture of objects - some examples mentioned in the meeting were pottery, glass, bricks, ceramics, pencils, clothing, metal items, paper, rubber, plastics, foods. Sources of information can be local firms, books, videos (or visits where already arranged). Some addresses are given at the back.

Encourage children to first of all think in groups about how an object was made and where it might have come from in the first place. Then even though they will be using secondary sources, try to ensure an interactive researching approach rather than passive absorption from those sources.

Guide children towards thinking about the various changes that might have taken place to raw materials in making the finished object. Let them show the changes by pictures as in a flow chart.

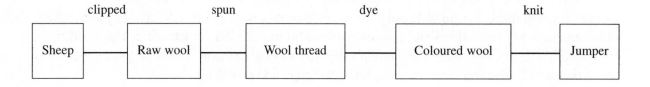

Product

A class display of flow-charts showing how various items have been made.

Activity 4 Finding Out More About Materials

a.   Finding out what materials are like

Choose questions which you feel your children could investigate in groups. They might raise questions themselves, e.g. 'Which piece of metal is the bendiest?' or 'Which materials can I see through?' or 'Is this piece of wood harder than this one?' You will have to select materials to match their investigation. Try to let them plan investigations themselves but be prepared to guide and advise them - encourage them, for example, to make their tests 'fair' (it may not be possible to have identical sizes of metal strips in testing for bendiness, for example, but they should still make it a fair comparison by hanging the same thing on each). See what ideas they come up with for answering the questions raised.

This activity is thus more structured with more guidance by the teacher than the similar more open investigation during the exploration phase.

Possible properties for investigation (according to the attainment target) are:

transparency, mass (weight), volume, strength,
hardness, solubility, flexibility, compressibility.

Product

Pupil accounts of their investigations (possibly as group records).   Where children cannot write, a brief report to the teacher might be noted down.

b.   Finding out how materials can be changed

Again materials will have to be selected to match questions raised - safety and suitability for the particular changes involved have to be borne in mind.

Possible changes to look at are:

by heating or cooking, e.g. jelly, pancakes

by mixing or dissolving, e.g. angel delight, egg whites

by pressing and rolling, e.g. for plasticine (this might be generalised through secondary sources to examples they cannot experience directly like pressing and rolling metal)

by bending and twisting, e.g. for art straws to make different shapes.

Product

Brief accounts of children's investigations (in the intervention diary).

DIARY OF INTERVENTION ACTIVITIES

NAME OF TEACHER _____

DATE	ACTIVITY

APPENDIX VII

POST-INTERVENTION INTERVIEW SCHEDULE

Types and Uses of Materials

═══════════════════

Aim 1. To ascertain any changes in children's ideas about solids, liquids and gases.

1.1   Use the containers A - E

      A = steel
      B = cotton wool
      C = treacle
      D = talcum powder
      E = vinegar

Ask: 1   Which do you think are solids?

           2   How did you decide?

           3   Which do you think are liquids?

           4   How did you decide?

           5   For any left over, as 'So what is this?'

1.2   Use the child's pictures of solids, liquids and gases
      (drawn after the intervention).

Ask: 6   Can you think of any more   (a) solids?
                                              (b) liquids?
                                              (c) gases?

           7   a.   What are solids like? How can you tell something is a solid?
                   b.   Repeat for liquid.
                   c.   Repeat for gas.

Aim 2. To find out children's ideas on the origins of materials and the possibility of their transformation.

2.1   Origins of materials

     8   Use the steel rod taken from A. Make sure children know what it is.

Ask: What do you think it was like before this? ... and before that? ... etc. (as far back as the children can go).

     9   Repeat for a piece of cotton thread (again make sure they are clear it is cotton).

2.2   Transformation of material

     10   Give the children the steel rod (A), piece of wire, piece of foil and filings.

Ask: Which can be changed into which?

     Make sure that all the possibilities which start with the rod are asked about.

     For those which can be changed (starting with the rod) ask how.

     For those which cannot, ask why not.

Aim 3. To see whether children can relate properties of materials to their use.

Ask: 11   Why do you think wood is a good material for making chairs?

     12   Why do you think metal is good material for making nails?

Aim 4:   To see if children understand the meaning of strong and hard and if they make tests fair.

13   Show two pieces of thread - cotton and wool.

How would you find out which one was stronger?

14   Show two pieces of metal.

How would you find out which one was harder?

APPENDIX VIII

BIBLIOGRAPHY

Andersson, B. (1984), *Chemical Reactions*, E.K.N.A. Group, University of Gothenburg, Gothenburg, Sweden.

Brook, A., Briggs, H. and Driver, R. (1984), *Aspects of Secondary Students' Understanding of the Particulate Nature of Matter*, Children's Learning in Science Project, University of Leeds, Leeds.

Driver, R. (1985), 'Beyond appearances: the conservation of matter under physical and chemical transformations' in Driver, R., Guesne, E. and Tiberghien, A. (eds), *Children's Ideas in Science*, Open University Press, Milton Keynes, pp. 145-69.

Jones, B. L., Lynch, P. P. and Reesink, C. (1989), 'Children's understanding of the notions of solid and liquid in relation to some common substances', *Int. J. Sci. Educ.*, Vol.11, No.4, pp. 417-27.

Modgil, S. (1974), *Piagetian Research. A Handbook of Recent Studies*, N.F.E.R., Windsor.

Novick, S. and Nussbaum, J. (1981), 'Pupil's understanding of the particulate nature of matter: a cross-age study', *Sci. Educ.*, Vol.65, pp. 187-96.

Osborne, R. J. and Cosgrove, M. M. (1983), 'Children's conceptions of the changes of state of water', *J. Res. Sci. Teaching*, Vol.20, No.9, pp. 825-38.

Piaget, J. (1930), *The Child's Conception of Physical Causality*, Kegan Paul, Trench Trubner and Co., London.

Piaget, J. and Inhelder, B. (1974), *The Child's Construction of Quantities*, Routledge and Kegan Paul, London.

Russell, T., Harlen, W. and Watt, D. (1989), 'Children's ideas about evaporation', *Int. J. Sci. Educ.*, Vol.11, Special Issue, pp. 566-76.

Russell, T. and Watt, D. (1990), *Evaporation and Condensation*, Primary SPACE Project Research Report, Liverpool University Press, Liverpool.

Séré, M. G. (1985), 'The gaseous state', in Driver, R., Guesne, E. and Tiberghien, A. (Eds), *Children's Ideas in Science*, Open University Press, Milton Keynes, pp. 105-23.

Stavy, R. (1990), 'Children's conception of changes in the state of matter: from liquid (or solid) to gas', *J. Res. Sci. Teaching*, Vol.27, No.3, pp. 247-66.

Stavy, R. and Stachel, D. (1985), 'Children's ideas about "solid" and "liquid"', *Eur. J. Sci. Educ.,* Vol.7, No.4, pp. 407-21.